Antioxidants in Nutrition,
Health, and Disease

Antioxidants in Nutrition, Health, and Disease

JOHN M. C. GUTTERIDGE

Oxygen Chemistry Laboratory, Critical Care Unit
Department of Anaesthesia & Intensive Care
Royal Brompton Hospital and National Heart & Lung Institute
London SW3 6NP, UK

and

BARRY HALLIWELL

Pharmacology Group, King's College
Manresa Road, London SW3 6LX, UK
and
Division of Pulmonary and Critical Care Medicine
University of California, Davis, Medical Center, CA 95817, USA

OXFORD UNIVERSITY PRESS

Oxford New York Tokyo

1994

Oxford University Press, Walton Street, Oxford OX2 6DP

Oxford New York Toronto
Delhi Bombay Calcutta Madras Karachi
Kuala Lumpur Singapore Hong Kong Tokyo
Nairobi Dar es Salaam Cape Town
Melbourne Auckland Madrid

and associated companies in
Berlin Ibadan

Oxford is a trade mark of Oxford University Press

Published in the United States
by Oxford University Press Inc., New York

A catalogue record for this book is available from the British Library

Library of Congress Cataloging in Publication Data
Gutteridge, John M. C.
Antioxidants in nutrition, health, and disease/
John M.C. Gutteridge and Barry Halliwell.
Includes bibliographical references.
1. Free radicals (Chemistry)—Pathophysiology.   2. Antioxidants—
Therapeutic use.   I. Halliwell, Barry.   II. Title.
[DNLM: 1. Antioxidants—therapeutic use.   2. Free Radicals.
3. Oxidants—adverse effects.   QV 800 G985a 1994]
RB170.G88   1994   616.07—dc20   94–18907

ISBN 0 19 854902 4 ✓

Typeset by Cambrian Typesetters
Frimley, Surrey
Printed in Great Britain
on acid-free paper by Bookcraft Ltd., Midsomer Norton, Avon

Preface

In this short text we attempt to describe the basic science behind free radicals, antioxidants, and their involvement in human nutrition, health, and disease. It is now well established that the increased formation of oxygen radicals and other oxygen derivatives frequently and, perhaps inevitably, accompanies tissue damage and degeneration. This observation has led to the widespread, but often incorrect, assumption that free radicals play a major causative role in human disease and in the ageing process. In recent years, there has been an explosion of interest about the use of 'antioxidant' nutritional supplements. Epidemiological evidence suggests that maintaining high intakes of some vitamins, minerals, and certain other food constituents may help to protect against life-threatening diseases such as heart disease and cancer. Is this effect due to antioxidant properties or to something else? Can nutritional supplements hope to cure or prevent many diseases? Should we all be taking vitamin E, vitamin C, or more of certain polyunsaturated fats?

The concise format of this book attempts to summarize, in a simple way, current thoughts about free radicals and antioxidants and to provide sufficient information for scientists, health professionals, and the lay public to develop their own opinions. The authors hope that the book will be useful to physicians, medical students, nurses, biologists, nutritionists, and even interested chemists (who may not be familiar with many of the biological aspects of free radicals).

In such a short text, it is impossible to cite all of the many distinguished scientists who have contributed to our knowledge of free radicals and antioxidants. Thus, we have provided only references to seminal papers and to major review articles.

London   J.M.C.G.
Davis   B.H.
August 1994

Acknowledgements

We are extremely grateful to the following scientists and publishers who granted us permission to reproduce their data:

The Upjohn Company, Dr David Hockley, Ann Dewar, Professor Randall Lauffer, Professor Hermann Esterbauer, Professor Antony Diplock, Professor Hermann Kappus, Dr A. Noronha-Dutra, Dr Thomas Eisner, Dr Daniel Aneshansley, Professor Bruce Ames, and Harwood Academic Press.

J.M.C.G. dedicates the book to his family: Puspha, Samantha, and Mark.

We are extremely grateful to Pat Wong (USA) and Jayne Wellington (UK) for their invaluable help with typing.

Contents

1 Oxygen, the breath of life or oxygen, the first toxic oxidizing air pollutant

A moralist, at least, may say that the air which nature has provided for us is as good as we deserve.

Joseph Priestley (1775)

Aerobic organisms are masters of the slow burn.

Ma and Eaton (1992)

The history of oxygen

The element oxygen (chemical symbol O) exists in air as a 'double' molecule, two atoms being joined together to give O_2 (dioxygen). Oxygen was first isolated and characterized between 1772 and 1774 by the individual skills of three great European scientists—Scheele, Priestley, and Lavoisier. Oxygen appeared in significant amounts in the Earth's atmosphere some 2.5×10^9 years ago, and geological evidence suggests that this was due to the photosynthetic activity of certain microorganisms, the blue-green algae. As they split water to obtain their essential requirement for hydrogen atoms, blue-greens released tonnes of oxygen into the atmosphere, creating perhaps the worst case of environmental pollution ever recorded on this planet.[†] The slow and steady rise in atmospheric oxygen concentrations was

---

[†] Many environmentalists worry that present-day pollution will 'destroy the planet'. Earth will probably survive, as it has survived many cataclysmic changes in its past (Table 1). It is the life forms on the Earth, including man, who might have a problem.

Table 1. *Some of the main events preceding the appearance of man on earth* (adapted from Harman, D. (1986). *Free radicals, ageing and degenerative diseases*. Alan R. Liss, Inc.)

Approximate time-scale (years ago)	
3.5 billion[†]	Intense solar radiation bombards the surface of the Earth. Free radical chemistry contributes to the formation of the first complex organic molecules. Anaerobic life begins, forming by-products such as sulphide, nitrite, and alcohols.
2.5 billion	Blue-green algae acquire the ability to split water and release O_2: $$2H_2O \rightarrow 4H + O_2 \uparrow$$
1.3 billion	Oxygen levels in the atmosphere reach 1%. Primitive anaerobic organisms disappear or retreat to oxygen-free areas. More-complex cells with nuclei (eukaryotes) begin to evolve. Eukaryotes and blue-greens develop into green leaf plants. Eukaryotes and prokaryotes able to reduce O_2 to H_2O eventually develop into animals. Emergence of multicellular organisms.
500 million	Oxygen levels in the atmosphere reach 10%. Ozone layer screens the land and allows life forms to emerge from the sea.
65 million	Primates appear.
5 million	Man appears. Atmospheric oxygen levels reach 21% (of dry air).

[†] Here one billion means one thousand million, not one million million as is sometimes used.

accompanied by the formation of the ozone layer in the strato-sphere. Both oxygen and the ozone layer acted as critical filters against the intense solar ultraviolet light striking the surface of the Earth. Ascending in altitude from the Earth's atmosphere into outer space we see a change from relatively heavy molecules such as O_2, N_2, and H_2O to lighter molecules, atoms and ions (such as hydrogen atoms and H^+ ions) and electrons, which are dominant above 800 km. The universe contains vast amounts of hydrogen (H) and helium (He). **Perhaps the Earth may be regarded as a unique centre of potential oxidation capacity in an otherwise reducing universe.**

The percentage of oxygen in air is now approximately 21 per cent. This makes oxygen the second most abundant element in the atmosphere. The first is nitrogen, N_2, at approximately 78 per cent. However, the actual mass of oxygen in the atmosphere is negligible when compared with that present as part of the water (H_2O) molecule in oceans, lakes, and rivers and that present as part of mineral ores (e.g. metal oxides such as the aluminium ore bauxite, Al_2O_3) in the Earth's crust, where oxygen is by far the most abundant element.

When the Earth's atmosphere changed from a highly reducing state to the oxygen-rich state that we know today (Table 1), the anaerobic life forms existing at that time either adapted, died, or retreated to places with little or no oxygen (Figure 1). Present-day anaerobic bacteria (which are killed by O_2 or cease growing when O_2

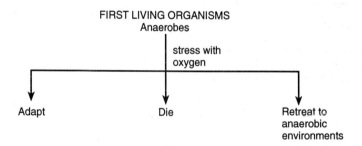

Figure 1. Adaptation to use O_2: (a) evolve antioxidant defences, (b) evolve O_2-using enzymes and electron transport chains, and (c) oxidize food material more efficiently, producing more energy per unit mass of food.

is present) are presumably the evolutionary descendants of organisms that followed this avoidance strategy. They are found in the mammalian colon, rotten foods, infected wounds, dental plaque, deep in soil, and similar places. However, none of these is a salubrious environment: in retrospect, organisms that evolved the capacity to *cope* with oxygen (by evolving antioxidant defence mechanisms) did best, because they could simultaneously evolve to *use* oxygen for efficient energy production and for other oxidation reactions. Thus every textbook of biochemistry explains how anaerobic metabolism of one molecule of the sugar glucose gives only two molecules of ATP (*adenosine triphosphate*, the energy currency of the cell), whereas aerobic (oxidative) metabolism gives 36 or 38 ATPs per glucose (depending on which textbook one reads). The evolution of efficient aerobic respiration allowed the development of complex multicellular organisms. However, since Earth's organisms consist mainly of water, and O_2 is poorly soluble in water (it is much more soluble in organic solvents), transport of the necessary O_2 to all the cells in a large organism required the evolution of high-capacity O_2 transporter molecules, of which the red-brown protein haemoglobin in our red blood cells is the prime example.

Aerobes use oxygen to oxidize ('burn') fuels rich in carbon and hydrogen (foods) to obtain the heat and other forms of energy essential for life. The carbon becomes carbon dioxide (CO_2) and the hydrogen becomes water (H_2O). Thus, the burning of glucose to provide energy in the body can be written as

$$C_6H_{12}O_6 + 6O_2 \rightarrow 6CO_2 + 6H_2O \qquad (1)$$

The carbon in glucose is oxidized to CO_2: the O_2 is reduced to water (these definitions are expanded in Figure 2). Fats can also be burned: they yield more energy per gram than do carbohydrates and so they are our major body food stores. Amino acids can also be burned: the nitrogen is first removed from them and eventually excreted in the urine as *urea*. The burning of dietary fuels in the cells produces ATP, a central energy source which is used to drive movement and the building up of cell components.

OXIDATION is   gain in oxygen:
                 $C + O_2 \rightarrow CO_2$ (carbon is oxidized to carbon dioxide)

                 or

                 loss of electrons:
                 $Na \rightarrow Na^+ + e^-$ (a sodium atom is oxidized to a
                                                 sodium ion)

                 $O_2^{\cdot-} \rightarrow O_2 + e^-$ (a superoxide radical is
                                                         oxidized to oxygen)

REDUCTION is   loss of oxygen:
                 $CO_2 + C \rightarrow 2CO$ (CO_2 is reduced to carbon monoxide;
                                                 C is oxidized to CO)

                 gain of hydrogen:
                 $C + 2H_2 \rightarrow CH_4$ (carbon is reduced to methane)

                 or

                 gain of electrons:
                 $Cl + e^- \rightarrow Cl^-$ (a chlorine atom is reduced to a
                                                 chloride ion)

                 $O_2 + e^- \rightarrow O_2^{\cdot-}$ (oxygen is reduced to superoxide
                                                         radical)

An OXIDIZING AGENT oxidizes another chemical by taking electrons from it, or by taking hydrogen, or by adding oxygen.

A REDUCING AGENT reduces another chemical by supplying electrons to it, by supplying hydrogen, or by removing oxygen.

Figure 2. Oxidation and reduction.

Some chemistry of oxygen

Oxygen is a stable, odourless, tasteless, and colourless gas. We have already pointed out its limited solubility in water. However, this limited solubility is vital to the survival of fish and other aquatic organisms, and essential for normal respiratory functions in man. Oxygen has to dissolve in water to cross the alveoli of the lungs and reach the transport protein haemoglobin. Oxygen solubility in water (like that of most gases) decreases with temperature—if the oceans and rivers get warmer, some fish and other aquatic organisms may not receive sufficient oxygen.

The air dissolved in water contains a higher percentage of oxygen (34 per cent) than does dry air (21 per cent), because O_2, although poorly soluble, is more soluble than is nitrogen. As already mentioned, oxygen is considerably more soluble in organic solvents than it is in water. For example, the fat solvent chloroform ($CHCl_3$) at 10 °C can dissolve up to 219.5 ml of oxygen per litre at one atmosphere pressure, whereas under the same conditions water dissolves only 38.2 ml of oxygen. These differences in solubility are important when considering the availability of oxygen for chemical reactions inside living systems: organic regions may contain more oxygen than aqueous regions. The membranes that separate cells from the environment and separate the different compartments within cells contain an interior organic phase, in which oxygen may tend to concentrate (see the Appendix to this chapter).

Why is O_2 toxic and what do antioxidant defences do?

The problem with oxygen is simply that it *oxidizes* organic molecules, including foods, plastics, paints, hydrocarbon[†] fuels (petroleum, gasoline, natural gas, etc.), rubber, and human tissues (the term oxidation is defined in Figure 2). Fortunately, the rates of these oxidations are very slow at normal temperatures. They can be increased by *enzymes* (such as those that catalyse oxidations of foodstuffs in aerobic cells) or by *heat*, as in combustion of hydrocarbon fuels. Antioxidant defences may prevent these oxidations by removing catalysts, by being preferentially oxidized (so protecting key cell components), or by repairing the damage caused by oxygen.

[†] A hydrocarbon contains carbon and hydrogen only. An example is the gas methane, CH_4.

Free radicals
What is a free radical?
The answer to the question of what is a free radical requires a brief review of some elementary chemistry. An atom consists of a central nucleus (containing positively charged protons and neutral neutrons) with electrons orbiting around it. The electrons associate in pairs, and each pair moves in its own region of space around the nucleus. **A free radical is defined as any species capable of independent existence (hence the term 'free') that contains one or more unpaired electrons.** The free radical nature of an atom or molecule is usually denoted by a superscript dot (e.g. $H^{\cdot}$, $O_2^{\cdot -}$, $OH^{\cdot}$). Molecules can be free radicals if one or more of the atoms present has unpaired electrons. Indeed, the diatomic oxygen molecule (O_2) qualifies as a free radical because it contains *two* unpaired electrons. Other gaseous free radicals include nitric oxide ($NO^{\cdot}$) and nitrogen dioxide ($NO_2^{\cdot}$). By contrast, the gas ozone (O_3) is not a free radical—there are no unpaired electrons in its molecules. Free radicals of different types vary widely in their chemical reactivity, but in general they are more reactive than nonradicals. Fortunately, however, oxygen is an exception to this rule: it is not a particularly reactive free radical because of peculiarities in its arrangement of electrons (the details need not concern us here).

When two free radicals meet, their unpaired electrons can join to form a pair, and both radicals are lost (Figure 3). However, since most molecules present in living organisms do not have unpaired electrons, any free radicals that are produced will most likely react with nonradicals, generating new free radicals. Hence, free radical reactions tend to proceed as *chain reactions* (Figure 3). This is a basic principle of free radical chemistry.

Unpaired electrons can be associated with many different atoms and molecules. Hence many different free radicals exist. For example, the structure of many proteins is stabilized by joining sulphur atoms together by covalent bonds[†] to form disulphide

---

[†] In a covalent bond, two atoms are chemically bonded together by sharing a pair of electrons between them.

(1)   ADDITION

$$x^• + y \rightarrow [x–y]^•$$

(2)   ELECTRON DONATION

$$x^• + y \rightarrow y^{•-} + x^+$$

(3)   ELECTRON REMOVAL

$$x^• + y \rightarrow x^- + y^{•+}$$

Only when two radicals meet can
chain reactions be terminated.

$$x^• + x^• \rightarrow x_2$$

$$x^• + y^• \rightarrow xy$$

Figure 3. Radicals beget radicals.

bridges. These disulphide bridges can be broken to give sulphur
radicals, if one of the two electrons from the covalent bond remains
on each sulphur atom

$$S - S \rightarrow S^• + S^• \tag{2}$$

Among the mechanisms by which these and other radicals can be
generated is the grinding of proteins (e.g. during the processing of
food material). The protein *keratin* present in human fingernails is
rich in disulphide bridges, and sulphur radicals can be detected in
fingernail clippings. However, in biological systems, most attention
has been focused on the oxygen radicals. Reduction of oxygen can
produce two free radicals—superoxide and hydroxyl (Figure 4).

Superoxide radical
Superoxide is made by adding one electron on to an oxygen
molecule. The added electron pairs with one of the two unpaired
electrons already present in O_2, leaving one unpaired electron. The

(1) $O_2 + e^- \longrightarrow O_2^{\cdot-}$ (superoxide radical)

(2) $O_2^{\cdot-} + 2H^+ + e^- \longrightarrow H_2O_2$ (hydrogen peroxide)

(3) $H_2O_2 + e^- \longrightarrow OH^- + OH^{\cdot}$ (hydroxyl radical)

(4) $OH^{\cdot} + e^- \longrightarrow OH^-$ (hydroxylion)

(5) $2OH^- + 2H^+ \rightarrow 2H_2O$

Overall $O_2 + 4H^+ + 4e^- \longrightarrow H_2O$

Figure 4. The reduction of oxygen (e^- is used to denote an electron).

chemical behaviour of $O_2^{\cdot-}$ differs greatly depending on what it is dissolved in. In water, $O_2^{\cdot-}$ is not very reactive. It can sometimes act as a weak oxidizing agent, by accepting one more electron. For example, it can oxidize ascorbic acid (vitamin C) (readers uncertain about oxidation and reduction can refresh their memories by consulting Figure 2).

$$\text{ascorbic acid} + O_2^{\cdot-} + H^+ \rightarrow \text{ascorbic acid radical} + H_2O_2 \quad (3)$$
hydrogen peroxide

Superoxide in aqueous solution more often acts as a reducing agent (e.g. it reduces ferric (Fe^{3+}) iron salts):

$$Fe^{3+} + O_2^{\cdot-} \rightarrow Fe^{2+} + O_2 \quad (4)$$

Upon donating its electron, $O_2^{\cdot-}$ is oxidized and re-forms dioxygen. However, in organic solvents, $O_2^{\cdot-}$ is far more reactive and dangerous. Thus any $O_2^{\cdot}$ generated in the interior of biological membranes (see the Appendix to this chapter) might do considerable damage.

Hydroxyl radical

The hydroxyl radical is the most reactive oxygen radical known to chemistry. It has tremendous potential for causing biological damage, since it attacks all biological molecules as soon as it comes into contact with them, usually setting off free radical chain reactions. Figure 5 shows an example of this. DNA contains a

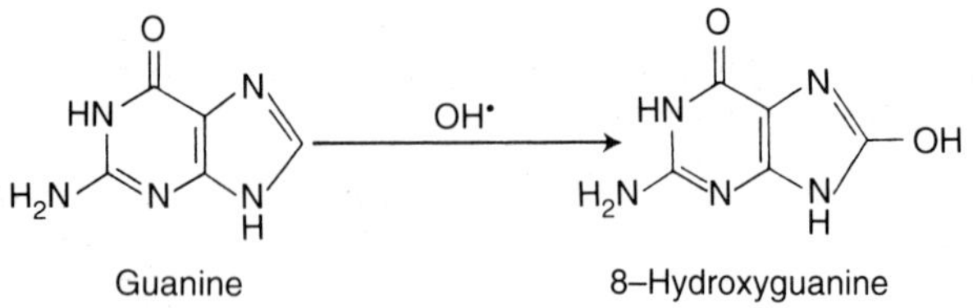

Figure 5. Reaction of hydroxyl radical with the body's genetic material, DNA.

backbone of the sugar *deoxyribose* linked by *phosphate* groups. Attached to deoxyribose are the *purine* bases *adenine* and *guanine* and the *pyrimidine* bases *cytosine* and *thymine*. Adenine pairs with thymine and guanine with cytosine. This pairing is essential in allowing DNA to copy itself when cells divide, and in reading the genetic information contained in DNA. Hydroxyl radicals can attack all constituents of DNA. The reaction in Figure 5 shows one of the things it can do to guanine. The OH˙ adds to guanine and makes a hydroxyguanine radical, which can then form 8-hydroxyguanine. 8-Hydroxyguanine can be misread when DNA is copied, introducing a *mutation*.

Most of our knowledge about the chemistry of OH˙ has been provided by radiation chemists, since exposing water to X-rays or γ-rays generates OH˙. Figure 6 explains how: it also explains the difference between the hydroxyl radical and the hydroxyl ion OH⁻, which are frequently confused in scientific writings. Hydroxyl ions are fairly harmless (unless one drinks a strong solution of the caustic alkalis sodium or potassium hydroxide, that is). Each litre of water in the human body at 37 °C contains about 6×10^{16} OH⁻ ions. If we had that number of hydroxyl *radicals*, we would never survive. Indeed, most of the damage done to living organisms exposed to excess ionizing radiation is caused by the consequences of OH˙ radical attack on biological molecules.

Hydroxyl radicals can add on to biological molecules, as illustrated by their reaction with guanine in Figure 5. They can also convert themselves back into water by pulling off hydrogen atoms (*hydrogen atom abstraction*) from a biological molecule, as happens in lipid

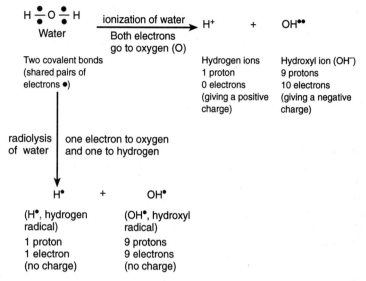

Figure 6. Water can be split into ions (with paired electrons, or no electrons at all) or free radicals (with unpaired electrons).

peroxidation (see Chapter 3). Since OH˙ reacts so fast with everything, any OH˙ generated combines with whatever is in the human body next to it: it doesn't last long enough to migrate anywhere else.

Hydrogen peroxide
If two hydroxyl radicals ever meet, they can join their unpaired electrons and make an oxygen–oxygen covalent bond, giving H_2O_2 (hydrogen peroxide), a product with no unpaired electrons

$$OH˙ + OH˙ \rightarrow H{-}O{-}O{-}H \tag{5}$$

Superoxide radicals generated in aqueous solution also make H_2O_2. One $O_2^{˙-}$ gives up its electron to another one. The first is oxidized to O_2, and the second is reduced to H_2O_2. The overall reaction is called the *dismutation* of $O_2^{˙-}$,

$$2O_2^{\cdot-} + 2H^+ \rightarrow H_2O_2 + O_2 \tag{6}$$

A few enzymes exist in the body that make H_2O_2 directly. For example, the amino acids incorporated into proteins by the human body are all L-type (this term describes the three-dimensional arrangement of their atoms in space). The alternative D-type amino acids are not used. However, they can be produced by bacteria in the gut and enter the body. These 'unnatural' D-amino acids are destroyed by a *D-amino acid oxidase* enzyme. The unwanted D-amino acids are oxidized, and the electrons removed from them are used to reduce O_2 to H_2O_2. Several human tissues (especially gut) contain the enzyme *xanthine oxidase* which oxidizes the complex molecule hypoxanthine to xanthine, and then oxidizes the xanthine further to uric acid. Uric acid is excreted in human urine: we do not have an enzyme to break it down further. When xanthine oxidase acts upon xanthine and hypoxanthine both $O_2^{\cdot-}$ and H_2O_2 are generated. The amounts of xanthine oxidase can increase as a result of tissue injury in certain human diseases, generating extra $O_2^{\cdot-}$ and H_2O_2 (see Chapter 6).

H_2O_2 is a pale-blue liquid, which mixes readily with water, diffusing easily in the body and crossing membranes. It has been used for many years as a bleach and to disinfect wounds. Since it has no unpaired electrons, it is not a free radical. However, if one extra electron is added to H_2O_2, it makes $OH^\cdot$ (Figure 4). **Hence H_2O_2 is a mobile time bomb:** it is poorly reactive itself, yet can make $OH^\cdot$ at any time if an electron is supplied to it. Where can this electron come from? One source is certain metal ions—those whose valencies differ by one. Iron is the prime example:

$$Fe^{2+} \rightarrow Fe^{3+} + e^- \tag{7}$$

ferrous   ferric
ion   ion

$$H_2O_2 + e^- \rightarrow OH^\cdot + OH^- \tag{8}$$

The net reaction is

$$H_2O_2 + Fe^{2+} \rightarrow OH^\cdot + OH^- + Fe^{3+} \tag{9}$$

Reaction (9) is the *Fenton reaction*. For many years, chemists have mixed iron(II) sulphate (ferrous sulphate, $FeSO_4$) and H_2O_2 to make $OH^{\cdot}$. Indeed, the Fenton reagent was first described in the 1890s by the Cambridge chemist H. J. H. Fenton, who showed that mixtures of ferrous salts and H_2O_2 could oxidize most organic molecules. Some titanium, copper, cobalt, and chromium salts also convert H_2O_2 into $OH^{\cdot}$.

Ultraviolet light and hydroxyl radicals
Ultraviolet (UV) light is not energetic enough to form $OH^{\cdot}$ from H_2O, but it can reverse equation (5). Hence UV irradiation of H_2O_2-containing systems can cause severe damage by making $OH^{\cdot}$.

Some other molecules to worry about
Oxides of nitrogen
Nitric oxide ($NO^{\cdot}$) and nitrogen dioxide ($NO_2^{\cdot}$) contain unpaired electrons and are therefore free radicals, whereas the 'laughing gas' nitrous oxide (N_2O) is not. Nitrogen dioxide is a dense brown poisonous gas and a powerful oxidizing agent. It is found in polluted air, smoke from burning organic materials, and at high levels in cigarette smoke. Nitric oxide, on the other hand, is a colourless gas and a weak reducing agent. Recently, biological interest in nitric oxide has centred around the observation that the cells lining our blood vessels (*vascular endothelial cells*), certain white blood cells that defend us against foreign organisms, and some cells in the brain actually produce nitric oxide, starting from the amino acid L-arginine. In blood vessels, $NO^{\cdot}$ relaxes muscles in the vessel wall, dilating the vessel and lowering blood pressure. It is sometimes called the 'endothelial-derived relaxing factor', or EDRF. Organic nitrites such as nitroglycerin and amyl nitrite (a common constituent of the recreational drug 'poppers') act by a similar mechanism. Although $NO^{\cdot}$ is biologically essential, production of excess $NO^{\cdot}$ in patients with severe infections can do harm (e.g. by lowering blood

pressure too much and contributing to septic shock). At body temperature, NO˙ can combine with oxygen to make $NO_2˙$,

$$2NO˙ + O_2 \rightarrow NO_2˙ \tag{10}$$

Among other reactions, $NO_2˙$ can pull off hydrogen atoms from membrane lipids and initiate lipid peroxidation (see Chapter 3).

$$NO_2˙ + lipid-H \rightarrow lipid˙ + HNO_2 \tag{11}$$

Hence, oxides of nitrogen are toxic in excess. Indeed, macrophages, cells that help defend us against foreign organisms (see the Appendix to Chapter 5) use oxides of nitrogen as one of the mechanisms by which they can kill parasites. *In small amounts*, oxides of nitrogen (at least NO˙) are useful. The same is true of many other free radicals, including $O_2˙^-$.

Ozone
The pale-blue gas ozone (O_3) provides an important protective shield in the stratrosphere against solar UV radiation. At ground level, however, ozone is an unwanted toxic substance, generated in polluted urban air by the action of light on some of the chemicals present. Ozone can also be produced by some scientific equipment and sometimes by photocopying machines. Ozone is irritating to the eyes, nose, and lungs. Indeed, it can oxidize and damage proteins, DNA, and lipids directly. These are *direct* oxidations; O_3 is not a free radical and does not itself start free radical chain reactions.

Singlet oxygen
Although oxygen O_2 has two unpaired electrons these electrons are arranged in such a way that O_2 oxidizes most things very slowly at room temperature. Indeed, one theory of ageing (the *free radical theory*), essentially states that *ageing is caused by the slow cumulative oxidation of body tissues over a lifetime.*

By a simple rearrangement of its electrons, it is possible to make O_2 much more reactive, converting it into *singlet oxygens*, which are powerful oxidizing agents. For example, O_2 oxidizes lipids at an immeasurably slow rate, but singlet oxygens oxidize them very fast into lipid peroxides (see Chapter 3). The rearrangement of electrons that produces singlet oxygen needs energy: the best-established source of this energy involves *photosensitization reactions*. When solutions of several coloured compounds are illuminated, they absorb energy from the light. Some of this absorbed energy can be transferred to O_2, making singlet O_2. This is a particular problem in green plants: they must be illuminated to drive photosynthesis and convert carbon dioxide (CO_2) from the air into sugars, but the green photosynthetic pigment chlorophyll can make singlet O_2. Compounds present in the human body that are capable of sensitizing singlet O_2 production include riboflavin (vitamin B_2) and haem. Of course, inside the body they are protected from light. However, bright sunlight can damage milk because milk contains riboflavin. The complex reactions of ozone with several biological molecules also make some singlet oxygen, and some singlet O_2 is produced during lipid peroxidation (see Chapter 3, Figure 3).

Several drugs (including some tranquilizers, antibiotics, and anti-inflammatory drugs) are photosensitizers and can cause skin damage if the patient sunbathes. Something similar occurs in the *porphyrias*, a series of disorders in which photosensitizing pigments accumulate in the skin. One striking example is *congenital erythropoietic porphyria*, fortunately a very rare disease. The condition presents at, or shortly after, birth with severe sensitivity to light. Extensive skin blistering with secondary infection leads to mutilation of exposed areas. The teeth (and bones) may be stained red due by the accumulated pigments and can glow in ultraviolet light. Excess body hair growth may occur. An interesting, if fanciful, theory is that these unfortunate individuals with hairy faces and shiny teeth who only ventured forth at night (to avoid the sun) gave rise to at least some of the *werewolf* legends. King George III of England may have suffered from porphyria, a disease that can be traced back through his ancestors James I and Mary Queen of Scots. The physicians of his day thought that he was suffering from 'evil humours', and treated him by blood-letting, using bloodsucking leeches.

Not all photosensitizing reactions are bad for us, however.

A mixture of porphyrin derivatives known as HPD ('haemato-porphyrin derivative') is used in cancer treatment. After injection of HPD, some tumours take it up and then become sensitive to damage if they are illuminated. This method is being used to treat some forms of lung cancer, in which the tumour can be illuminated with light from a tube placed down the throat.

Appendix to Chapter 1
Cholesterol, saturated, and unsaturated fats. What do they do in the body?

Membranes encircle all cells and control the exchange of material between the cell and its surroundings. Inside cells, membranes separate the different cell compartments (*organelles*, such as nucleus and mitochondria) from the fluid cell matrix (*the cytosol*).

The main constituents of membranes are lipid and protein, the amount of protein increasing with the number of functions the membrane performs. 'Lipid' is a general term used to describe any biological compound that is soluble in organic solvents such as chloroform and ether. The term includes both molecules that contain fatty acids, examples being *triglycerides* and *phospholipids*, and molecules containing hydrocarbon ring structures, examples being cholesterol, steroid hormones, and some of the fat-soluble vitamins.

Fatty acid structure: saturates and polyunsaturates

Until recently, the term polyunsaturated was used by only a handful of lipid biochemists and nutritionists. Following the introduction and promotion of polyunsaturated margarines, however, the word has become very familiar to physicians and to the general public and has often been associated with the concept of 'healthy fats'. More recently, a healthy image for oils rich in polyunsaturated fats, such as *evening primrose oil* and *fish oils*, has been promoted by extensive advertising. Yet others have cautioned that polyunsaturated fats are prone to attack by free radicals, generating toxic products. Let us explore the background to these assertions.

All fatty acids[†] contain a chain of carbon atoms with attached hydrogen, terminating in a carboxyl (−COOH) group. The hydrocarbon chain gives the fatty acid its solubility in organic solvents, whilst the −COOH group is acidic, being able to generate H^+ ions:

$$-COOH \rightleftarrows -COO^- + H^+ \tag{A1}$$

If all the carbon atoms in the side chain are joined by single covalent bonds, the fatty acid is said to be *saturated*. Thus palmitic acid, $CH_3(CH_2)_{14}COOH$, and stearic acid, $CH_3(CH_2)_{16}COOH$, are saturated. The shorthand $C_{16:0}$ and $C_{18:0}$ is often used to denote them in the scientific literature: 16 and 18 carbon atoms present, respectively, and 0 carbon–carbon double bonds. Double covalent bonds between carbon atoms ($C=C$) are more reactive than single ones. If one double bond is present, the acid is *monounsaturated*: oleic acid ($C_{18:1}$) is an example (Table A1). The double bond links carbon atoms 9 and 10 in oleic acid. Double bonds can exist in two geometrical shapes, *cis* and *trans*. As Figure A1 shows, the *cis* form of oleic acid has a kinked chain, whereas the *trans* form does not. In the human body, *cis* forms are much more common than *trans* forms. Because the kinks in the chains make it harder for the molecules to pack together, *cis* fatty acids need lower temperatures to make them solidify than do *trans* ones, that is they have lower melting points. Saturated fats have higher melting points because the molecules can pack together more easily and most of them are solid at room temperature. By adding hydrogen to the double bonds in unsaturated fats, saturated fatty acids can be generated. This is done by the process of *catalytic hydrogenation*, used in the manufacture of some solid margarines from liquid vegetable oils containing unsaturated fats. By careful control of reaction conditions, oils can be hydrogenated to achieve a product with the desired consistency: hydrogenation is not usually taken to completion, since the product would often be too hard. However, partial hydrogenation of unsaturated fatty acids can lead to production of *trans* fatty acid forms. **Debate rages in the nutrition literature as to whether *trans* fatty acids are harmful or not, but there is no clear answer as yet.**

[†] An *acid* is a donor of protons, H^+ ions. H^+ ions have one proton and no electrons.

Table 1. *Some common naturally occurring fatty acids*

Shorthand name	Common name	Occurrence
16:0	Palmitic	Natural fats and oils, especially palm oil
18:0	Stearic	Natural fats and oils, especially beef fat
18:1 (n−9)*	Oleic	Natural fats and oils, especially olive oil
18:2 (n−6)	Linoleic	Widespread, many seed oils
18:3 (n−3)	α-Linolenic	All plant leaves, some seed oils, e.g. soybean, rapeseed, linseed oils
20:4 (n−6)	Arachidonic	Animal membranes
20:5 (n−3)	Eicosapentaenoic	Fish oils
22:6 (n−3)	Docosahexaenoic[†]	Fish oils, nervous system

* The numbering system in parentheses identifies double bonds from the end of the chain farthest from the carboxyl (i.e. the methyl, $-CH_3$, end).
[†] Docosahexaenoic acid is particularly important in the human brain and in the retina of the eye.

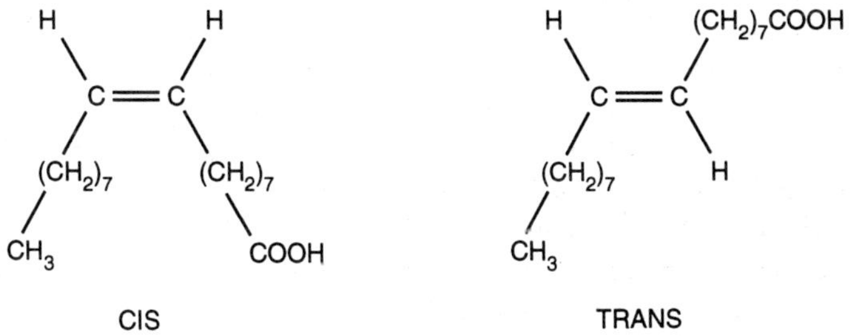

Figure A1. The *cis*- and *trans*-forms of oleic acid, $C_{18:1}$ (note the kink in the chain of the *cis* form).

The *polyunsaturated* fatty acids contain two or more double bonds. Perhaps the best known in humans is *arachidonic acid* ($C_{20:4}$), which can be synthesized in the body from linoleic acid ($C_{18:2}$) in the diet. α-Linolenic and linoleic acids are thought to be essential in the human diet (the *essential fatty acids*), and are mainly obtained from foods of vegetable origin (Table A1).

Triglycerides: in the diet and in body fat stores

Some fatty acids exist free in the human body (e.g. transported in blood, attached to the plasma[†] protein *albumin*), but the great majority are in the form of *lipids*, of which several classes exist. *Triglycerides* are used as a long-term store of energy in our adipose (fat) tissue. They represent most of the fat in the diet. Most food triglycerides contain a mixture of fatty acids: saturated, mono-unsaturated, and polyunsaturated. Despite statements in the popular press, there are no pure 'saturated' or 'unsaturated' sources of dietary fat. All that differs is the *relative composition* of fatty acid side chains present. Thus, although coconut oil, for example, is commonly referred to as a saturated fat, it contains some linoleic acid ($C_{18:2}$). Similarly, to speak of corn oil as a polyunsaturated fat obscures the fact that approximately 20 per cent of its fatty acid side chains are saturated. Lard has almost equal amounts of saturated and monounsaturated fats.

In a triglyceride, three fatty acids are linked by covalent bonds to an alcohol called *glycerol*. The three fatty acid side chains are not randomly distributed in most natural oils and fats: there is a tendency for particular fatty acids to be located at specific positions on the three glycerol carbons. The same is true of fatty acids in positions 1 and 2 of *phospholipids*, a second major class of lipids. Phospholipids make up the bulk of cell membranes in humans (Figure A2). The fatty acid side chains of the phospholipid molecule (R_1 and R) are *hydrophobic* (water hating) whereas the choline

---

[†] Blood plasma is the straw-coloured fluid left after the red and white blood cells are removed. It contains many proteins.

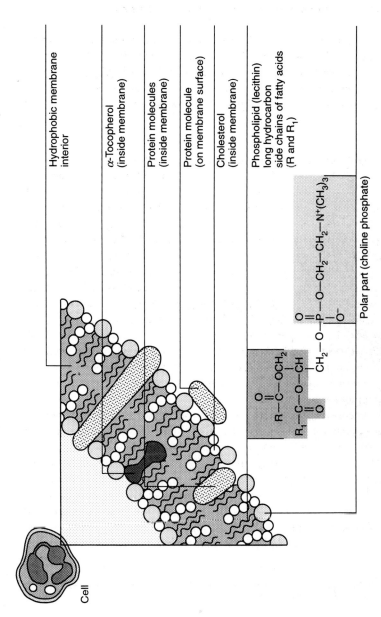

Hydrophobic membrane interior

α-Tocopherol (inside membrane)

Protein molecules (inside membrane)

Protein molecule (on membrane surface)

Cholesterol (inside membrane)

Phospholipid (lecithin) long hydrocarbon side chains of fatty acids (R and R_1)

Cell

Polar part (choline phosphate)

$$R_1-C-O-CH$$
$$O=C-OCH_2$$
$$CH_2-O-P-O-CH_2-CH_2-N^+(CH_3)_3$$
$$O=P-O^-$$

Figure A2. The structure of membranes. (Courtesy of Upjohn Company 'Current Concepts' series.)

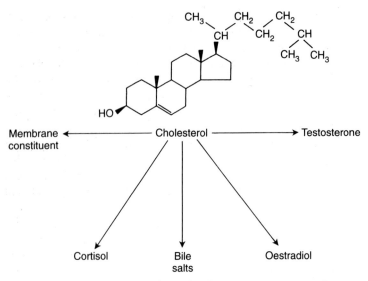

Figure A3. The structure of cholesterol.

phosphate part of the molecule is *hydrophilic* (water loving). Membranes have a *bilayer structure*, which generates a hydrophobic interior (two rows of fatty acid side chains) and hydrophilic areas in contact with the aqueous phase on either side of the cell membrane. Proteins are present on the outside and within the membrane. The alcohol glycerol is linked to two fatty acid side chains (R_1 and R) and to a phosphate group and a nitrogen-containing compound. *Lecithin* is a common phospholipid in human membranes: the nitrogen compound present is *choline*.

Cholesterol

The lipid *cholesterol* (Figure A3) is present in membranes and blood lipoproteins†. An adult human contains about 2 g cholesterol per

---

† A *lipoprotein* contains lipid and protein. Human blood contains several types of lipoprotein: this is discussed in detail in Chapter 3.

kilogram of body weight. In lipoproteins, fatty acids are often attached to cholesterol to give *cholesterol esters*. Cholesterol is often supposed to be a 'baddie', yet it is essential for human life. It helps to maintain membrane structure, it is used to make the *bile salts* (secreted into the gut by the liver, to help break up dietary fats for ease of digestion) and the *steroid hormones* such as oestrogens (female sex hormones), testosterone (male sex hormone), and cortisol, another hormone essential to body function. **Indeed, if the diet contains insufficient cholesterol, the liver makes it from other ingested fats.** It is then distributed to body tissues using lipoproteins in the blood as carriers. **That is why even drastic decreases in dietary cholesterol intake often produce only small drops in blood cholesterol levels.**

Biological membranes essentially consist of a bilayer of lipid (usually phospholipid; Figure A2) in which the hydrophilic heads point outward towards the watery fluids on the outside and inside of the cell, whereas the hydrophobic hydrocarbon fatty acid 'tails' point inwards to produce a hydrophobic membrane interior. Proteins are embedded in the bilayer to perform specific functions, such as to act as transporter molecules (carrying substances across the membrane) or to catalyse chemical reactions on the membrane surface (enzymes). To allow its correct functioning, the membrane must be 'fluid', that is its constituents must be able to move around freely. Fluidity is largely determined by the presence of poly-unsaturated fatty acid side chains in the membrane lipids. The downside to this is that membrane polyunsaturated fatty acid side chains are very susceptible to free radical attack, which starts the process of *lipid peroxidation* (Chapter 3).

2 Metals and oxygen: respiration, oxidation, and oxygen toxicity

Gold is for the mistress—silver for the maid—
Copper for the craftsman cunning at his trade!
'Good!' said the Baron, sitting in his hall,
'But iron'—Cold Iron—is master of them all.

Rudyard Kipling

Aerobic respiration

Aerobic respiration is the process by which O_2 is used to oxidize foodstuffs (fats, carbohydrates, proteins) and produce heat and other forms of energy such as ATP. Because O_2 is poorly soluble in water, the major constituent of the human body[†], we have evolved a transport protein to carry it, haemoglobin in the red blood cells (Figure 1). Haemoglobin has four protein subunits. To each is bound a complex chemical structure (*haem*), at the centre of which is an iron ion, in the ferrous (Fe^{2+}) state. Oxygen reversibly binds to this iron, giving a Fe^{2+}–O_2 complex (causing the bright red colour of oxygenated blood). In fact, there is a slight shifting of electrons between the iron and oxygen, so that this oxyhaemoglobin is intermediate between a Fe^{2+}–O_2 (ferrous–oxygen) complex and an

---

[†] Approximately 70% of the adult human body is water!

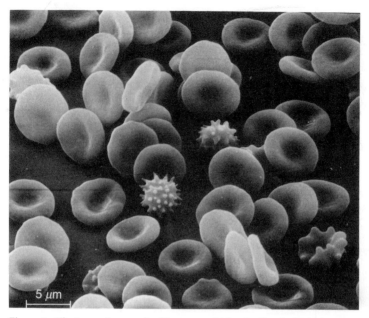

Figure 1. Electron micrograph of human red blood cells of which there are around five thousand million per millilitre of blood. The red cells contain the protein haemoglobin which carries oxygen. (Courtesy of Dr David Hockley.)

$Fe^{3+}-O_2^{\cdot-}$ (ferric superoxide)[‡] complex, although it is closer to the former.

Haemoglobin makes superoxide radicals

Oxyhaemoglobin slowly releases superoxide radical ($O_2^{\cdot-}$) and forms ferric haemoglobin, often called *methaemoglobin*. Methaemoglobin is unable to bind and transport O_2, and is dark brown in colour. In the human body, this release of $O_2^{\cdot-}$ may happen only once in a thousand cycles of O_2 binding and release, but there is a

[‡] Remember that superoxide radical, $O_2^{\cdot-}$, is the one-electron reduction product of oxygen (Chapter 1).

large mass of haemoglobin in the body and so total body $O_2^{\cdot-}$ production by this mechanism is significant (it is estimated that up to 3 per cent of our haemoglobin releases $O_2^{\cdot-}$ in this way every day). Hence red blood cells need antioxidant defences to protect themselves against $O_2^{\cdot-}$, plus an enzyme to reduce Fe^{3+} in methaemoglobin back to the functional Fe^{2+}-form (*methaemoglobin reductase*).

Mitochondria reduce oxygen and make energy

About 85–90 per cent of the O_2 that we breathe in is utilized by the mitochondria; these organelles are the major source of ATP in the human body. Essentially, food materials are oxidized: they lose electrons, which are accepted by electron carriers (Figure 2), such as nicotinamide adenine dinucleotide (NAD^+) and flavins (FMN and FAD). The resulting reduced nicotinamide adenine dinucleotide

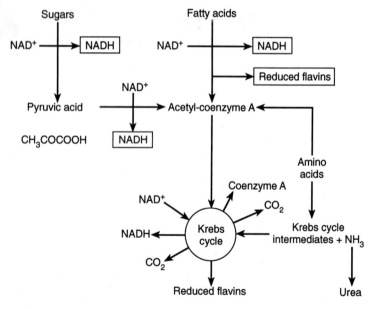

Figure 2. Basic human metabolism-energy production.

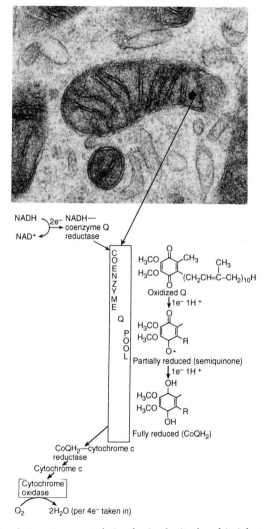

Figure 3. The electron transport chain of animal mitochondria (photograph courtesy of Ann Dewar). Mitochondria have an outer membrane and an infolded inner membrane which contains the electron transport chain. The figure shows mitochondria in lung.

(NADH) and reduced flavins ($FMNH_2$ and $FADH_2$) are re-oxidized by oxygen in mitochondria, producing large amounts of energy. Oxidation is catalysed in a stepwise fashion, so that the energy is released gradually, by the *electron transport chain* present in the inner mitochondrial membranes (Figure 3). Electrons pass from NADH to 'non-haem iron proteins'—these accept the electrons by converting their bound iron from Fe^{3+} to Fe^{2+}. They pass on the electrons by re-oxidizing to Fe^{3+}. Later in the chain, the *cytochrome proteins* work in the same way, except that their iron ions are bound to haem rings. Cytochromes accept electrons by forming Fe^{2+}–haem, and release them again by oxidizing to Fe^{3+}–haem. (In this respect, they are very different from the haem protein haemoglobin which only works in the Fe^{2+} form). Figure 3 shows, in a simplified form, the other constituents of the chain.

The part of the electron transport chain that actually uses O_2 is the terminal oxidase enzyme, *cytochrome oxidase*. It removes one electron from each of four reduced (Fe^{2+}–haem) cytochrome c molecules, oxidizing them to ferric cytochrome c. It adds the four electrons on to O_2; the overall reaction is

$$O_2 + 4H^+ + 4e^- \rightarrow 2H_2O \tag{1}$$

However, it is chemically impossible to add four electrons to O_2 at once—it must be done in stages. Hence cytochrome oxidase is a very complex enzyme, because it catalyses several reduction steps. Since partially reduced oxygen species are damaging, the enzyme must also keep them safely bound to its active sites unit they are fully converted to water (Equation 1).

Cytochrome oxidase has both iron and copper ions bound to it. These metals play key roles in oxygen reduction and safe binding of partially reduced oxygen intermediates. Cytochrome oxidase has a very high affinity for O_2: it still works well at O_2 concentrations of less than 1 millimetre of mercury (mmHg).[†] Hence energy production by mitochondria can continue at low O_2 concentrations. This is

---

[†] Gas pressure is measured in terms of the length of a column of mercury (Hg) the gas can support. Dry air (21% O_2, 78% N_2) at 0 °C can support 760 mm Hg.

important—fully oxygenated blood leaving the lungs has an O_2 concentration of some 100 mmHg, but this falls rapidly in the tissues as oxyhaemoglobin unloads O_2. Cells deep within a tissue may experience extracellular O_2 concentrations of only 5–15 mmHg, and O_2 still has to cross the cell to reach mitochondria. Hence the O_2 concentration within actively respiring mitochondria may be very low. Perhaps this is one antioxidant defence.

Some enzymes use oxygen directly and make radicals

The 10–15 per cent of O_2 we breathe in that is not taken up by mitochondria is used by various *oxidase* and *oxygenase* enzymes, and also by direct chemical ('non-enzymic') reactions. Thus D-*amino acid oxidase* uses O_2 to oxidize unwanted D-amino acids, making H_2O_2 (Chapter 1). *Xanthine oxidase* oxidizes xanthine and hypoxanthine into uric acid, making superoxide radical ($O_2^{•-}$) and H_2O_2 (Chapter 1). The *cytochromes P450* are enzymes found in many body tissues (especially concentrated in liver). They use O_2 to oxidize a wide range of 'foreign compounds' (drugs, toxins, pesticides, etc.) into products that are, usually, less toxic; that is, cytochromes P450 are a detoxification system. In human tissues, cells are embedded in a *matrix* containing, among many other compounds, the protein *collagen* which gives strength and flexibility. When collagen is being made in the body *proline and lysine hydroxylase* enzymes use O_2 to put essential hydroxyl ($-OH$) groups on to the amino acids proline and lysine in the protein. Synthesis of the essential hormones *epinephrine* (adrenaline) and *norepinephrine* (noradrenaline) involves oxygen-dependent addition of $-OH$ groups to the amino acid tyrosine by another hydroxylase enzyme. In general, these O_2-using enzymes bind O_2 much less efficiently than cytochrome oxidase does. Hence oxidases and hydroxylases at normal cellular O_2 concentrations, are often limited in their action by O_2 concentration.

Several oxidase and oxygenase enzymes produce potentially dangerous species. Thus D-amino acid oxidases make H_2O_2, and xanthine oxidase makes both $O_2^{•-}$ and H_2O_2. Some of the other enzymes can make $O_2^{•-}$ as a minor product. For example, certain cytochromes P450 'leak' some electrons on to O_2 during their functioning and make $O_2^{\cdot-}$. If more O_2 is around, then more $O_2^{\cdot-}$ and H_2O_2 will usually be made, since these systems are rarely saturated with O_2.

Mitochondria make radicals

A small percentage of the electrons passing down the mitochondrial electron transport chain (Figure 3) also leak from the correct path directly on to O_2, making $O_2^{\cdot-}$: this leakage tends to happen in the early part of the chain (before the cytochromes). Again, the rate of leakage increases with O_2 concentration. Thus giving mitochondria more O_2 will not necessarily speed up electron transport (because of the very high affinity of cytochrome oxidase for O_2), but it will cause more electron leakage and $O_2^{\cdot-}$ formation. Fortunately, cytochrome oxidase, although it makes dangerous partially reduced forms of O_2, does not release them and does not leak electrons.

Some body constituents react directly with oxygen and make radicals

Some direct chemical reactions between O_2 and body constituents can occur. For example, a solution of the hormone epinephrine (adrenaline) deteriorates on exposure to air—it oxidizes easily. Epinephrine reacts with O_2 to make $O_2^{\cdot-}$, which then oxidizes more of the hormone in a free radical chain reaction. The human body contains many other molecules which behave similarly—they are often called 'auto-oxidizable molecules' because they can react with O_2 to make $O_2^{\cdot-}$. Indeed, oxyhaemoglobin could be classified under this heading, since it can form $O_2^{\cdot-}$ and methaemoglobin

$$\text{haemoglobin} - Fe^{2+} - O_2 \rightarrow \text{haemoglobin} - Fe^{3+} + O_2^{\cdot-} \quad (2)$$

Metals: useful but dangerous

Oxygen-using enzymes and electron transport chain components (Figure 3) contain metals, often iron and sometimes copper (Table 1).

Table 1. *A few of the important iron- and copper-containing proteins involved in oxygen metabolism*

Metalloprotein	Metals	Function(s)
Superoxide dismutases	Copper, manganese, iron	Antioxidant defence (see Chapter 3)
Catalase	Iron (haem ring)	Antioxidant defence (see Chapter 3)
Peroxidases	Iron (haem ring)	Use H_2O_2 to oxidize molecules
Cytochromes	Iron (haem ring)	Electron transport
Haemoglobin	Iron (haem ring)	Oxygen transport
Myoglobin	Iron (haem ring)	Oxygen storage in 'red' muscles and heart
Cytochrome oxidase	Iron, copper	Terminal oxidase of the mitochondrial electron transport chain
Ferritin	Iron	Iron storage
Transferrin	Iron	Iron transport
Lactoferrin	Iron	Iron binding
Caeruloplasmin	Copper	Can load iron on to transferrin and into ferritin; supplies copper to cells; antioxidant defence
Cytochromes P450	Iron (haem ring)	Detoxification of foreign compounds
Proline/lysine hydroxylases	Iron	Collagen formation
Dopamine-β-hydroxylase	Copper	Hormone synthesis

These metals catalyse oxidation reactions because they can transfer single electrons by changing their valency state, for example

$$Fe^{2+} \rightleftarrows Fe^{3+} + e^- \tag{3}$$

as in cytochromes, and

$$Cu^+ \rightleftarrows Cu^{2+} + e^- \tag{4}$$

as in cytochrome oxidase. Metals also play a key role in 'autoxidation' reactions. For example, the reactions of epinephrine and ascorbic acid (vitamin C) with O_2 are very slow unless traces of iron or copper ions are present. Indeed, they might not react at all in the absence of metals; it is impossible to remove all traces of iron and copper from laboratory chemicals, since these metals are ubiquitous in the environment. Thus 'free' iron and copper in the human body can be dangerous because they will accelerate 'autoxidation' reactions: they can also do this in foodstuffs. For example, the presence of traces of copper (e.g. from copper cookware) can rapidly destroy vitamin C in foods. Metals with a fixed valency, such as zinc (Zn^{2+}), aluminium (Al^{3+}), and magnesium (Mg^{2+}), are never used in enzymes to catalyse oxidation reactions because they cannot mediate electron transfers. However, magnesium and zinc are used in many other types of enzyme.

Iron and copper: important enzyme-bound catalysts

Iron is essential to aerobic life. It is the fourth most abundant element on the Earth. Within the physiological pH range[†], Fe^{3+} is insoluble in water. The ion Fe^{2+} is more soluble, but readily oxidizes to Fe^{3+}

$$O_2 + Fe^{2+} \rightarrow Fe^{3+} + O_2^{\cdot-} \tag{5}$$

How quickly it does so depends on what the Fe^{2+} is attached to. For example, the protein and haem ring in haemoglobin help to keep

---

[†] pH is a measure of the degree of acidity of a solution, defined as $-\log_{10}$ of the hydrogen ion (H^+) concentration. Because it is a *negative* log, pH values *fall* as acidity *rises*. pH values <7 are acidic, >7 are alkaline. The pH of most human body fluids and cells is usually close to 7.4, that is slightly alkaline.

Fe^{2+} reduces oxygen:

$$Fe^{2+} + O_2 \longrightarrow Fe^{2+}\text{--}O_2 \longrightarrow Fe^{3+} + O_2^{\cdot-}$$

Superoxide makes H_2O_2:

$$2O_2^{\cdot-} + 2H^+ \longrightarrow H_2O_2 + O_2$$

H_2O_2 reacts with Fe^{2+} in a Fenton reaction:

$$H_2O_2 + Fe^{2+} \longrightarrow OH^\cdot + OH^- + Fe^{3+}$$

This makes highly damaging hydroxyl radicals (OH·).

(handwritten margin notes:) $O_2 + e^- \to O_2^{\cdot-}$ $2\,O_2^{\cdot-} + 2H^+ \to H_2O_2 +$

Figure 4. Why Fe^{2+} can cause biological damage.

the iron as Fe^{2+}. By contrast, high concentrations of simple Fe^{2+} salts can damage biological materials under aerobic conditions. Figure 4 shows why.

Since the atmosphere of the early Earth was reducing, iron may have existed as Fe^{2+}. As increasing levels of O_2 appeared, Fe^{2+} probably became oxidized to Fe^{3+}, which is insoluble. Deposits of iron ores nearly always contain ferric iron. Yet organisms still need to take in iron because of the essential roles that it plays (Table 1). To overcome this problem, microorganisms adapted by synthesizing iron-chelators[†] (*siderophores*). These were able to solubilize and capture ferric ions and carry them inside the cells to provide the iron required. Today, microbial siderophores, such as *desferrioxamine*, are among some of the most effective iron chelators used to treat iron-overload diseases in humans (see later in this chapter).

In Chapter 1, we said that oxygen is unable to undergo easy reaction with covalent molecules, making it sluggish in most of its chemical reactions. Our definition of a free radical allows us to include Fe^{2+} and Fe^{3+} ions as radicals, because they possess unpaired electrons. Their variable valency, allowing addition or loss of a single electron, makes iron and copper powerful promoters of O_2-dependent reactions. From the list of proteins shown in Table 1, we can see how important iron has become as a biological catalyst of

---

[†] *Chelator* comes from an ancient word meaning 'claw' (chela). A chelator 'grasps' the iron atom by binding to it at least two points.

aerobic metabolism. Copper is also important, although used less often than iron: the body of an average human contains about 4.5 g iron but only about 0.08 g copper.

Iron and copper are dangerous catalysts if 'free'

Iron and copper at the active sites of enzymes are excellent selective catalysts[†] of oxidation reactions (Table 1). One problem is that the 'free' metal ions are also good catalysts, and this can lead to biological damage. Thus free iron and copper accelerate autoxidation reactions, and both can react with H_2O_2 to form highly dangerous OH·

$$Fe^{2+} + H_2O_2 \rightarrow Fe^{3+} + OH^· + OH^- \tag{6}$$

$$Cu^+ + H_2O_2 \rightarrow Cu^{2+} + OH^· + OH^- \tag{7}$$

Superoxide ions can re-reduce the oxidized metal ions (Fe^{3+}, Cu^{2+})

$$Fe^{3+} + O_2^{·-} \rightleftharpoons Fe^{2+} + O_2 \tag{8}$$

$$Cu^{2+} + O_2^{·-} \rightarrow Cu^+ + O_2 \tag{9}$$

Reaction (8) is reversible (see reaction (5)). If we combine equations (6)–(9), we arrive at

$$O_2^{·-} + H_2O_2 \xrightarrow{Fe, Cu} O_2 + OH^· + OH^- \tag{10}$$

that is, iron and copper ions catalyse a reaction of $O_2^{·-}$ with H_2O_2 to form OH·. Thus the poorly reactive (at least in aqueous solution) H_2O_2 and $O_2^{·-}$ can form the devastatingly reactive OH· if they encounter free iron or copper ions.

Aerobic organisms minimize this problem, because they have evolved methods to absorb iron and copper from the gut, transport them in the blood, and deliver them to cells: these methods also prevent or minimize the catalytic ability of these metal ions. **Iron and copper are simply *not allowed* to be free in the human body unless absolutely unavoidable—they are almost always bound to carrier proteins or locked away in storage proteins.**

[†] A catalyst accelerates a chemical reaction but is not itself used up. For example, in equation (10) Fe^{2+} and Cu^+ form OH· from H_2O_2 and are recycled by $O_2^{·-}$ (equations 8 and 9). Thus they *catalyse* reaction (10).

Human iron metabolism

An average adult human male contains some 4.5 g iron, absorbs about 1 mg iron per day, and when in iron balance excretes the same amount. Since the total iron turnover is around 35 mg per day, the body must have extremely efficient mechanisms for preserving absorbed body iron. Only slight disturbances to this delicate balance between iron intake and iron loss can push the body into conditions of iron overload or iron deficiency. It has been estimated that in the world today, some 500 million people are iron deficient and several million are iron overloaded. No specific mechanisms exist for iron excretion, loss occurring by the shedding of cells lining the gut, in sweat, faeces, urine, and by menstrual bleeding in women.

Most of the body's iron (some two-thirds) is found in haemoglobin, with smaller amounts present in the muscle protein myoglobin, various enzymes (Table 1), the iron transport proteins transferrin and lactoferrin, and the iron storage proteins ferritin and haemosiderin. Myoglobin acts as a 'reservoir' of O_2 in red muscles and in the heart. Like haemoglobin, it contains a haem ring and binds O_2. It holds on to this O_2 tightly, but can release it again if the supply of O_2 to the tissue becomes abnormally low.

Most of our dietary intake of iron is in the form of non-haem ferric ion. This requires solubilization and reduction before absorption from the intestine can occur. Hydrochloric acid in the gastric juice within the stomach and ascorbic acid (vitamin C) aid absorption by solubilizing the iron and reducing it to the Fe^{2+} state. Dietary haem-iron is mostly derived from red meat products and can be directly absorbed by the intestine. Only a small, and carefully controlled, fraction of the total iron ingested is absorbed and allowed to enter the circulation. In the inherited disease *idiopathic haemochromatosis* (IH), this regulation fails and too much iron enters the body. The time taken for clinically significant iron overload to develop is often 40 or more years, depending on the diet to some extent. Eventually, transferrin becomes completely saturated with iron and the blood contains free iron (actually probably loosely bound to albumin and various small molecules). The liver removes this free iron rapidly, until it becomes heavily iron-overloaded and unable to absorb any more. The consequences of prolonged iron overload include:

- liver damage
- diabetes
- heart malfunction
- joint inflammation
- liver cancer

Treatment is by copious bloodletting, to deplete the excess body iron as fast as possible.

Any unwanted iron is stored by cells in the protein ferritin. Iron within ferritin is not available to catalyse free radical reactions. There are several reports that superoxide radical ($O_2^{\bullet-}$) can release iron from ferritin: this does not seem to be iron in the core, but iron associated with the protein coat, and only a tiny percentage of the iron in ferritin can be released by $O_2^{\bullet-}$. Thus ferritin is far safer than an equivalent amount of free iron. Nevertheless, an excess of $O_2^{\bullet-}$ might interact with ferritin to release a little iron and provide the essential ingredients for $OH^{\bullet}$ generation (Figure 4). If large amounts of iron need to be stored, ferritin can be converted into another protein, *haemosiderin*. Since haemosiderin is insoluble in water, it presumably is a safe 'non-catalytic' form of iron.

Copper metabolism

Whenever possible, iron is safely bound to proteins in the human body. The same is true for copper. Unlike the abundant iron, copper is a trace element in the Earth's crust, and some sixty times less abundant than iron in the human body. Copper salts are usually more reactive than iron salts in oxygen radical reactions (e.g. they react faster with H_2O_2 to make $OH^{\bullet}$), perhaps one reason why less copper is used. Absorption of dietary copper takes place in the stomach, and upper small intestine, from where it is probably transported to the liver bound to the plasma protein albumin. Human albumin has one site that binds copper very tightly, plus a number of weaker binding sites. The liver takes copper off albumin and incorporates most of it into the protein ceruloplasmin, a large protein containing 6 or 7 copper ions per molecule, which is then released into the plasma. Ceruloplasmin accounts for at least 90 per cent of the total copper present in human blood plasma.

Most of the copper attached to ceruloplasmin will not catalyse free radical reactions. This protein may act as a donor of copper to certain cells. It can also oxidize Fe^{2+} and facilitate loading of the resulting Fe^{3+} on to transferrin or into ferritin. Whereas non-enyzmic oxidation of Fe^{2+} generates damaging free radicals (Figure 4), ceruloplasmin-catalysed oxidation does not.

Copper bound to albumin can still interact with H_2O_2 to give $OH^{\cdot}$, but this radical then attacks the albumin itself and does not enter free solution to damage other molecules. Although the albumin will

Table 2. *Some consequences of exposure to elevated oxygen concentrations*

Animal used	Mode of exposure	Observation
Rat	5 atm pure O_2	Convulsive paralysis
Rat	100% O_2, 3 days	Severe lung damage
Adult humans	Breathing pure O_2, >6 h	Alveolar damage
Premature babies	$O_2 > 21\%$ in incubators	Increased risk of damage to the retina of the eye to an extent that can cause blindness (*retrolental fibroplasia*, sometimes called *retinopathy of prematurity*)
Hamsters	70% O_2, 3–4 weeks	Damage to the testes
Guinea-pigs	70% O_2, 6–36 days	Inhibition of red cell formation in bone marrow

undergo free radical damage as a result of such reactions, it can be replaced easily (there are 30–50 mg/ml of albumin in plasma and the liver replaces it continuously). By directing any copper-dependent free radical damage on to itself, albumin may protect more important biological molecules during the transport of copper from the gut to the liver.

Damage by oxygen: the phenomenon of oxygen toxicity

Although aerobes have adapted to life in an atmosphere of 21 per cent O_2, excess O_2 damages them. Table 2 gives some examples. Perhaps even 21 per cent O_2 exerts slow cumulative damaging effects, which contribute to the process of ageing (see Chapter 6). The phenomenon of 'oxygen toxicity' was originally attributed to direct inhibitory effects of O_2 itself upon enzymes. Although isolated examples of this do occur in living systems, it was soon realized that direct inactivation of enzymes by O_2 cannot account for most of the damaging effects of elevated O_2 concentrations.

Gerschmann and Gilbert in the USA were the first to propose (in 1954) that the damaging effects of oxygen are caused by oxygen radicals. Their theory was later expanded by McCord and Fridovich (also in the USA), following their discovery of the superoxide dismutase[†] enzymes, into a 'superoxide theory of oxygen toxicity'. Superoxide dismutases (SODs) are important biological antioxidant defence enzymes, apparently specific for the catalytic removal of $O_2^{\cdot-}$ radicals. Considerable evidence shows that SODs are very important physiological antioxidants (see Chapter 3). It follows that removal of $O_2^{\cdot-}$ *in vivo* is an important process.

The superoxide theory of O_2 toxicity proposes that excess O_2 is damaging because it causes formation of too much $O_2^{\cdot-}$ (e.g. by increasing leakage of electrons on to O_2 from electron transport chains and cytochromes P450) and by increasing the activity of oxidase enzymes such as xanthine oxidase, which are not working as fast as they can at normal cellular O_2 concentrations. The theory

[†] Chapter 3 gives an account of the superoxide dismutases.

is supported by a great deal of experimental evidence, but how does $O_2^{\cdot-}$ cause damage? Is the problem $O_2^{\cdot-}$ itself? Scientists have shown in bacteria that some enzymes, essential in metabolism, can be directly inactivated by $O_2^{\cdot-}$. Perhaps $O_2^{\cdot-}$ might do the same in human cells, but this has not yet been demonstrated. Superoxide accelerates $OH^\cdot$ production from H_2O_2 in the presence of iron and copper ions (equation 10) and SOD inhibits. Indeed, an excess of $O_2^{\cdot-}$ can mobilize limited amounts of Fe^{2+} from ferritin and provide all the necessary ingredients for $OH^\cdot$ generation (Figure 4). Another source of $OH^\cdot$ might be the reaction of $O_2^{\cdot-}$ with nitric oxide, the endothelium-derived relaxing factor (see Chapter 1).

$$O_2^{\cdot-} + NO^\cdot \rightarrow \underset{\text{peroxynitrite}}{ONOO^-} \tag{11}$$

$$ONOO^- + H^+ \rightarrow ONOOH \tag{12}$$

$$ONOOH \rightarrow OH^\cdot + NO_2^\cdot \tag{13}$$

All these processes *may* be important, but we are not yet sure exactly what is bad about too much $O_2^{\cdot-}$ in the human body.

3 Antioxidants: elixirs of life or media hype?

'Too many free radicals, that's your problem.'
'Free radicals, Sir?'
'Yes. They're toxins that destroy the body and brain—caused by eating too much red meat and white bread and too many dry Martinis.'
'Then I shall cut out the white bread, Sir.'

James Bond, *Never Say Never Again* (1983)

Introduction

The first organisms to evolve on the Earth were anaerobes: there was initially little or no O_2 in the atmosphere. As the O_2 content rose, some organisms adapted by evolving antioxidant defence systems to protect themselves against the toxic effects of O_2 (see Chapter 1). The term 'antioxidant' is widely used in newspapers, magazines, and the scientific literature, but it is rarely defined. Often it is used to refer only to antioxidants than can scavenge radicals, such as superoxide dismutase, vitamin E, and ascorbic acid (vitamin C). Let us explore this further.

Antioxidants

Aerobic organisms have evolved many types of antioxidant defence. Indeed, the authors define an antioxidant as **any substance that delays or inhibits, oxidative damage[†] to a target molecule.** All

[†] *Oxidative damage* is a broad term used to cover the attack upon biological molecules of oxygen-containing free radicals such as superoxide ($O_2^{\cdot-}$) and hydroxyl ($OH^{\cdot}$), and also attack by nonradical oxygen derivatives such as singlet oxygens and ozone.

molecules present in living organisms are potential targets of oxidative damage: lipids, proteins, nucleic acids, and carbohydrates. When antioxidants are being studied in the laboratory, a target of attack must be selected. It may be chosen because it is important. It is often chosen merely because damage to it is easy to measure. For example, damage to lipids (lipid peroxidation) can be caused by free radical attack: it is often measured because simple assays exist to measure it. However, other cell components can be attacked by radicals, and damage to them could be equally or more important. Thus laboratory studies of 'antioxidants' must be interpreted with close attention to the methods used. Saying that 'X is a good antioxidant' is meaningless without specifying the methods used. From first principles, it is easy to see that antioxidants might protect a target by:

(1) scavenging oxygen-derived species, either by using protein catalysts (enzymes) or by direct chemical reaction (in which case the antioxidant will be used up as the reaction proceeds);
(2) minimizing the formation of oxygen-derived species;
(3) binding metal ions needed to convert poorly reactive species (such as $O_2^{\cdot-}$ and H_2O_2) into nasty ones (such as $OH^{\cdot}$);
(4) repairing damage to the target;
(5) destroying badly damaged target molecules and replacing them with new ones.

The human body uses all of these mechanisms.

Protection of cells

Cells have a formidable armamentarium of defences, because they make free radicals and other oxygen-derived species continuously. Examples we have already discussed are:

(1) formation of $O_2^{\cdot-}$ by electron leakage from mitochondria and cytochromes P450, by autoxidation reactions, and by the action of certain enzymes (such as xanthine oxidase);
(2) formation of H_2O_2 by dismutation of $O_2^{\cdot-}$ and by the action of certain oxidase enzymes;

(3) formation of OH˙ from 'background' ionizing radiation—we are constantly exposed to low-level radiation, which presumably splits some water to make OH˙ in the body.

Within cells, metals such as iron and copper have to be removed from transport proteins and conveyed to their sites of final use within the cell. An example is the removal of iron from transferrin and its insertion into cytochromes in mitochondria. It is widely speculated that the intracellular 'transit pools' of iron and copper ions exist in a form that could catalyse conversion of $O_2^{˙-}$ and H_2O_2 into OH˙, although almost no hard data exist because it is extremely difficult to measure these transit pools. Assuming their existence, however, it may be seen that cells have a particular interest in removing $O_2^{˙-}$ and H_2O_2, before they can contact metal ions and make damaging OH˙.

The first line of antioxidant defence in cells is provided by enzymes.

Superoxide dismutases

It was the discovery of the superoxide dismutase (SOD) enzymes, first described in 1968 by McCord and Fridovich in the USA, that converted the 'free radical theory of oxygen toxicity' into a 'superoxide theory of O_2 toxicity' (see Chapter 2). SODs are present in almost all aerobic organisms (Table 1) and aerobes deficient in them are unable to grow properly in the presence of O_2. For example, if the bacterium *E. coli*[†] lacks its normal SOD activity, it suffers membrane damage, inability to synthesize certain amino acids, and greatly increased mutation rates under aerobic conditions. Indeed, a massive amount of evidence indicates that superoxide dismutases are important components of human antioxidant defence.

SODs contain metals essential for their catalytic function—it is ironic that copper and iron, if free, are powerful promoters of

---

[†] *Escherichia coli* is a bacterium found in the human colon. It has been widely used for laboratory studies. Many different strains exist.

Table 1. *Types of superoxide dismutase (SOD): all catalyse the reaction* $2O_2^{\cdot-} + 2H^+ \rightarrow H_2O_2 + O_2$

Metals present within the SOD proteins	Organisms containing them	Subcellular location in humans
Copper and zinc: the copper is the 'catalytic' metal, zinc helps to maintain the enzyme structure. This form of the enzyme is denoted Cu,Zn-SOD	Almost all aerobic eukaryotes (cells with nuclei), some bacteria	Cytosol, nucleus, possibly some in peroxisomes (small organelles containing a range of oxidase enzymes, together with catalase to destroy H_2O_2)
Manganese [Mn-SOD]	Many bacteria, most aerobic eukaryotes	Mitochondria, sometimes in the cytosol as well
Iron [Fe-SOD]	Many bacteria, a few higher plants	Not present in humans

oxidative damage but that the body needs them not only because of their key roles in O_2 transport, metabolism, and respiration (see Chapter 2) but also because they are required for antioxidant defence systems. The metals bound to SOD catalyse the reaction of two $O_2^{\cdot-}$ molecules with H^+ ions to form H_2O_2 and O_2. This reaction occurs slowly at pH 7.4 (see Chapter 2),

$$2O_2^{\cdot-} + 2H^+ \rightarrow H_2O_2 + O_2 \tag{1}$$

but SODs accelerate it by 10 000 times.

In humans, the manganese-containing SOD (Mn-SOD) in mitochondria presumably removes $O_2^{\cdot-}$ produced as a result of electron leakage on to O_2 from the mitochondrial electron transport chain and by mitochondrial oxidase enzymes. The copper- and zinc-containing SOD (Table 1) Cu,Zn-SOD deals with $O_2^{\cdot-}$ from cytosolic oxidases and from the cytochrome P450 enzymes, which are located in the *endoplasmic reticulum*[†] of the cell. Some Cu,Zn-SOD may be present in the nucleus and some is present in *peroxisomes*. These are small organelles, bounded by a single membrane, that contain a range of oxidase enzymes, such as D-amino acid oxidase, glycolate oxidase, and some of the enzymes involved in the oxidation of fatty acids to make energy in the human body.

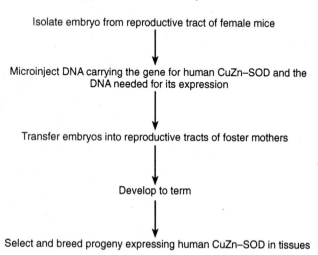

Isolate embryo from reproductive tract of female mice

⬇

Microinject DNA carrying the gene for human CuZn–SOD and the DNA needed for its expression

⬇

Transfer embryos into reproductive tracts of foster mothers

⬇

Develop to term

⬇

Select and breed progeny expressing human CuZn–SOD in tissues

The resulting mice have elevated levels of CuZn–SOD and:

(i)   are more resistant than controls to O_2 toxicity,

(ii)   are more resistant than controls to certain toxins,

(iii)   show abnormal neuromuscular junctions in the tongue,

(iv)   may show some of the other neurological defects characteristic of Down's syndrome.

Figure 1. Principles of generating mice transgenic for human CuZn-SOD.

[†] The endoplasmic reticulum is an internal membrane network present in most cells. Among other functions, it handles proteins destined for export from the cell and contains the cytochrome P450 enzymes, which detoxify foreign compounds.

The non-enzymic breakdown of $O_2^{\cdot-}$ makes H_2O_2 (equation 1) and the SOD enzymes speed this up. Thus SODs must work in conjunction with enzymes that destroy H_2O_2. If left to itself, H_2O_2 could migrate from its site of formation and make $OH^\cdot$ when it contacted iron or copper ions. H_2O_2 also poses a threat to Cu,Zn-SOD itself; if H_2O_2 accumulates, it can inactivate the enzyme. Experiments in which the levels of antioxidant defence enzymes are manipulated in mammalian cells grown in the laboratory illustrate this point clearly: too much SOD, in relation to the activities of H_2O_2-metabolizing enzymes, can damage the cells.

Such imbalance might also be medically relevant. In *Down's syndrome* (Figure 1), the most frequent genetic defect is that three copies of chromosome 21 are present in the affected child. The gene encoding Cu,Zn-SOD lies on this chromosome, and so affected individuals have about 50 per cent more Cu,Zn-SOD activity than usual in the tissues. Debate has raged as to whether the excess of Cu,Zn-SOD causes or contributes to the disorder. Experiments involving the construction of *transgenic animals* are contributing to this debate (Figure 1) and early results suggest that the excess of Cu,Zn-SOD does indeed contribute. Thus, the activities of anti-oxidant defence enzymes must be balanced.

Catalase

Catalase is a large enzyme, containing haem-bound iron at its active sites[†]. In most mammalian tissues, catalase is located in small organelles called *peroxisomes*, although in heart tissue there *may* be some catalase in the mitochondria as well. Catalase removes H_2O_2 by breaking it down directly into O_2:

$$2H_2O_2 \; \rightarrow 2H_2O + O_2 \tag{2}$$

Catalase has an enormous capacity to destroy H_2O_2: in terms of molecules of H_2O_2 destroyed per minute per molecule of enzyme, it is one of the most active enzymes known. However, its affinity for

[†] The *active site* of an enzyme is the part of the molecule which actually catalyses the reaction and to which the *substrate* of the enzyme (i.e. the molecule it acts upon) binds.

H_2O_2 is also low—thus it needs high H_2O_2 concentrations to work fast. Putting it another way, catalase deals only slowly with H_2O_2 at low concentrations.

Peroxidases

Peroxidases are enzymes that *use* H_2O_2 to oxidize a substrate. If SH_2 is used to denote the substrate, the general reaction catalysed by peroxidases may be written as

$$SH_2 + H_2O_2 \rightarrow S + 2H_2O \tag{3}$$

Plants and bacteria frequently contain peroxidases that oxidize a very wide range of molecules. Hence they are often called 'non-specific' peroxidases. *Horseradish peroxidase* is one of the most studied peroxidase enzymes in biology—it can use H_2O_2 to oxidize almost anything in the laboratory, although what it actually does oxidize inside the horseradish plant is unknown. It is rare to find nonspecific peroxidases in human tissues, although they are present in some phagocytic cells (see Chapter 4). Nonspecific peroxidases are often claimed to be present in the uterus and stomach, but this may be due to the phagocytes normally resident in these tissues.

Nonspecific peroxidases are haem-containing proteins, which usually have much higher affinity for H_2O_2 than does catalase. However, there are chemical similarities between catalase and peroxidase: indeed catalase has a weak peroxidase activity on a few substrates, such as ethyl alcohol (the alcohol of alcoholic drinks). However, this plays little, if any, role in alcohol metabolism in humans, as there is a much more efficient alcohol-oxidizing enzyme (*alcohol dehydrogenase*) in the liver.

Glutathione and glutathione peroxidases

Human tissues contain *glutathione peroxidases* as their major peroxide-removing enzymes. These enzymes remove H_2O_2 at a high rate by using it to oxidize reduced glutathione (GSH) into oxidized glutathione (GSSG)

$$2GSH + H_2O_2 \rightarrow GSSG + 2H_2O \tag{4}$$

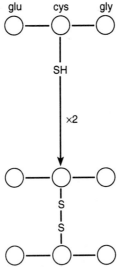

Figure 2. Structure of glutathione. Reduced glutathione (GSH) is made of 3 amino acids joined together (a tripeptide). Oxidized glutathione (GSSG) is made by joining two reduced glutathione molecules by their −SH groups, losing the two hydrogens, and forming a disulphide bridge.

As Figure 2 explains, GSH is a tripeptide with a free thiol (−SH) group, found in mammalian cells at high concentrations. The GSSG must then be converted back to GSH: that is the function of *glutathione reductase*, an enzyme containing FAD (a derivative of the water-soluble vitamin riboflavin, vitamin B_2)

$$GSSG + NADPH + H^+ \rightarrow NADP^+ + 2GSH \qquad (5)$$

NADPH (a molecule resembling NADH) is used by glutathione reductase as a source of reducing power.

GSH has antioxidant properties in addition to its use by glutathione peroxidases. For example, $OH^•$ can attack DNA at certain sites by abstracting hydrogen atoms to give DNA radicals ($DNA^•$)

$$DNA + OH^• \rightarrow DNA^• + H_2O \qquad (6)$$

Radiation chemists frequently claim that GSH can repair this damage by replacing the hydrogen. This has been the basis for developing compounds containing thiol (−SH) groups as *radio-protective agents*, to help people accidentally exposed to excess radiation

$$DNA^{\cdot} + GSH \rightarrow DNA + GS^{\cdot} \tag{7}$$

$$GS^{\cdot} + GS^{\cdot} \quad \rightarrow GSSG \tag{8}$$

Reaction (7) makes GS$^{\cdot}$, a glutathione radical (GS$^{\cdot}$). This radical is usually said to join with another such radical to make GSSG (equation 8), although some scientists believe that the GS$^{\cdot}$ radical might react with O_2 to make other, more damaging, radicals.

GSH has many other roles in the human body. For example, many toxins (such as chlorobenzenes) are metabolized to safer products by being combined with GSH in reactions catalysed by the *glutathione transferase* enzymes, which are widespread in mammalian cells. This detoxification system is additional to the cytochromes P450.

The glutathione peroxidase that metabolizes H_2O_2 (equation 4) is largely present in the cytosol of mammalian cells, but some is also in the mitochondria. These are the same locations as SOD, suggesting that glutathione peroxidase is the main enzyme that deals with H_2O_2 produced by Cu,Zn-SOD in the cytosol and Mn-SOD in the mitochondria.

Selenium
Glutathione peroxidase is very unusual in that it requires *selenium* for its activity. Selenium is a chemical element that has some resemblance to sulphur in its properties. Trace amounts of selenium are needed in the diet, although large amounts are poisonous. Selenium may perform several roles in the human body. For example, it appears to be involved in the synthesis of hormones by the thyroid gland, but its most important role in the human body is as an essential constituent of the glutathione peroxidase enzymes.

Protection of cell membranes and other lipids
Catalase, GSH, glutathione peroxidases, glutathione reductase, and

the SOD enzymes appear to reside in the aqueous part of the cell. What about the lipid part of the cell?

Human cell membranes consist of a phospholipid bilayer (see Chapter 1, Figure A1). Proteins are embedded in the bilayer to perform specific functions, such as transport molecules and enzymes. To function properly, the membrane must be 'fluid' (i.e. its constituents must be able to move around freely). Fluidity is largely determined by the presence of polyunsaturated fatty acid side chains in the membrane lipids (see Chapter 1, Figure A3). The downside to this is that polyunsaturated fatty acid side chains are very susceptible to free radical attack, which can start off *lipid peroxidation* (Figure 3). Peroxidation causes damage to membrane lipids and proteins, and depletion of antioxidants. Reactive free radicals can pull off hydrogen atoms from polyunsaturated fatty acid side chains. A hydrogen atom (H$^\cdot$) has only one electron. This hydrogen is bonded to a carbon in the fatty acid backbone by a covalent bond (a shared pair of electrons). Hence, the carbon from which H$^\cdot$ is abstracted now has an unpaired electron (i.e it is a free radical):

$$C-H + \text{reactive radical}^\cdot \rightarrow C^\cdot + \text{radical}-H \qquad (9)$$

Polyunsaturated fatty acid side chains (two or more double bonds) are much more easily attacked by radicals than are saturated (no double bonds) or monounsaturated (one double bond) side chains.

The hydroxyl radical is one species that can initiate lipid peroxidation:

$$C-H + OH^\cdot \rightarrow C^\cdot + H_2O \qquad (10)$$

Singlet O_2 can react directly with fatty acids to produce peroxides (see Chapter 1).

When C$^\cdot$ radicals are generated in the hydrophobic interior of membranes (equation 9), their most likely fate is combination with O_2 dissolved in the membrane (O_2 is much more soluble in organic solvents that it is in water):

$$\text{lipid}^\cdot + O_2 \rightarrow \underset{\text{peroxyl radical}}{\text{lipid}-O_2}{}^\cdot \qquad (11)$$

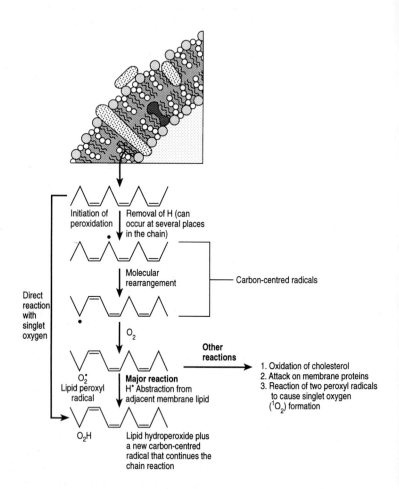

Figure 3. Lipid peroxidation. Peroxidation of a fatty-acid side chain with three double bonds is shown. Removal of an atom of hydrogen, the initiating event, can occur at several different places in the side chain. (Courtesy of Upjohn Co, 'Current Concepts' series.)

The resulting peroxyl radical is highly reactive: it can attack membrane proteins and oxidize adjacent polyunsaturated fatty acid side chains (Figure 3)

$$lipid-O_2{}^{\cdot} + lipid-H \rightarrow lipid-O_2H + lipid^{\cdot} \tag{12}$$

Reaction (11) is repeated and the whole process continues in a *free radical chain reaction*. The overall reaction is

$$lipid + O_2 \rightarrow lipid-O_2H + \text{damaged membrane proteins} \tag{13}$$

Lipid$-O_2$H, *lipid hydroperoxides*, are more hydrophilic than unperoxidized fatty acid side chains. They try to migrate to the membrane surface to interact with water, thus disrupting the membrane structure, altering fluidity and making the membrane leaky.

Protection of membranes is achieved by three mechanisms: radical scavenging, lipid repair, and lipid replacement.

Radical scavenging: tocopherols and tocotrienols

The major scavenger inside human membranes is *d-α-tocopherol*, often called vitamin E. The term *vitamin E* was first used to refer to a fat-soluble 'factor' discovered in 1922 to be essential in the diet of rats to permit normal reproduction. In nature, eight substances have been found to have vitamin E activity: d-α-, d-β-, d-γ-, and d-δ-tocopherols, and d-α, d-β, d-γ, and d-δ-tocotrienols. The most effective form found in membranes is d-α-tocopherol, often called *RRR-α-tocopherol* (Figure 4). The name 'tocopherol' comes from the Greek words *tokos* (childbirth) and *phero* (to bring forth). The other tocopherols are less important in humans because, although they can act as antioxidants, they are absorbed less well from the gut and are retained less well in the body tissues, making them less effective overall (Figure 4). Synthetic 'vitamin E' (dl-α-tocopherol, sometimes called All-*rac*-α-tocopherol) contains about 12.5 per cent of d-α-tocopherol, together with seven other tocopherols that are less biologically active (21–90 per cent).

Tocopherols inhibit lipid peroxidation because they scavenge lipid peroxyl radicals (equations 11 and 12) much faster than these

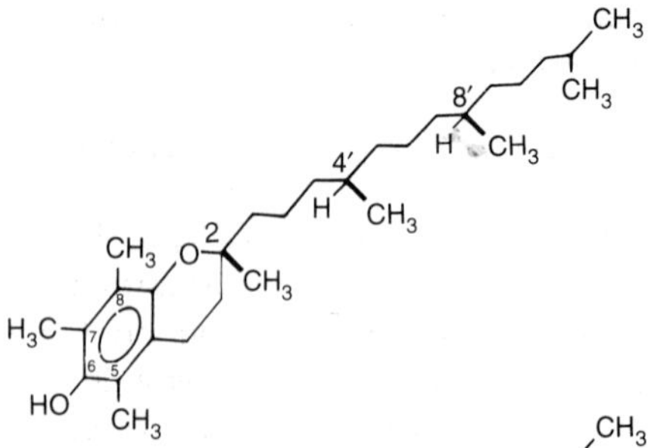

RRR-α-tocopherol

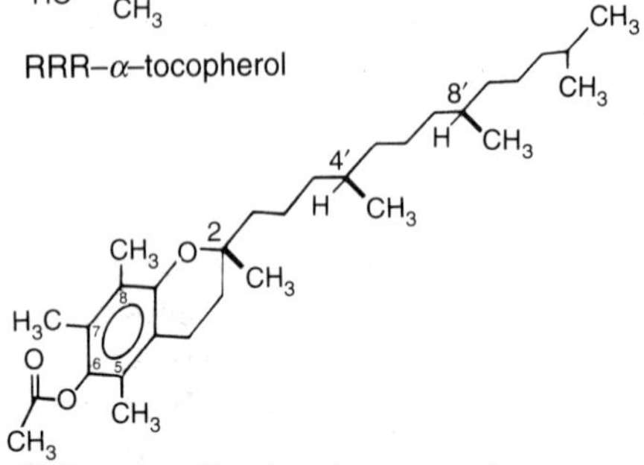

RRR-α-tocopheryl acetate

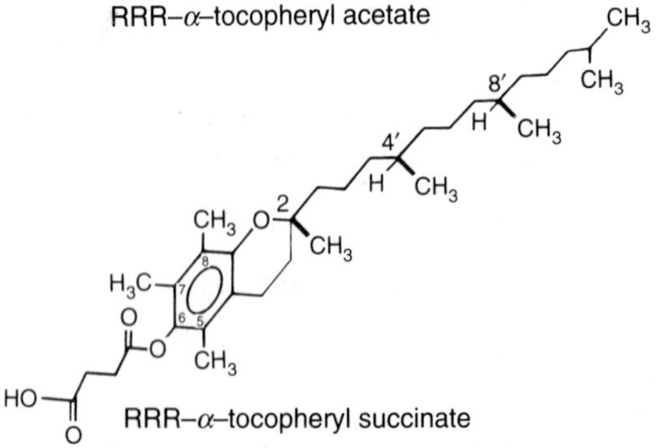

RRR-α-tocopheryl succinate

Figure 4. Structural formulae of natural vitamin E in its free form (RRR-α-tocopherol or d-α-tocopherol) and its esters with acetic acid (RRR-α-tocopheryl acetate) and succinic acid (RRR-α-tocopheryl succinate). The esters are often used in commercial vitamin E preparations because they are more stable than vitamin E itself. (From 'Tolerance and Safety of Vitamin E,' by H. Kappus and A.T. Diplock, with thanks to these authors and to the Vitamin E Research and Information Service.)

radicals can react with adjacent fatty acid side chains or with membrane proteins:

$$\alpha\text{-tocopherol} + \text{lipid}-O_2^{\cdot} \rightarrow \alpha\text{-tocopherol}^{\cdot} + \text{lipid}-O_2H \quad (14)$$

The $-OH$ group of the tocopherol (Figure 4) gives up its hydrogen atom to the peroxyl radical, converting it to a lipid peroxide. This, of course, leaves an unpaired electron on the $O^{\cdot}$ to produce a tocopherol radical. Hence α-tocopherol is a *chain-breaking antioxidant*—it breaks the chain reaction of lipid peroxidation but is itself converted to a radical during the process (equation (14)), that is the vitamin E is consumed.

Radical scavenging: Ubiquinol (reduced coenzyme Q)

Another compound present in membranes that might sometimes act as a chain-breaking antioxidant is the reduced form of coenzyme Q, ubiquinol. Coenzyme Q plays an important role in mitochondrial electron transport (see Chapter 2), but it may also act as an antioxidant. This is presumably why coenzyme Q is now appearing alongside vitamin E on the shelves of health food shops. There is a debate in the scientific literature as to whether ubiquinol in the human body scavenges peroxyl radicals itself:

$$\text{lipid}-O_2^{\cdot} + \text{CoQH}_2 \rightarrow \text{lipid}-O_2H + \text{CoQH}^{\cdot} \quad (15)$$

or whether it acts by regenerating α-tocopherol:

$$\alpha\text{-tocopherol}^{\cdot} + \text{CoQH}_2 \rightarrow \alpha\text{-tocopherol} + \text{CoQH}^{\cdot} \quad (16)$$

It might, of course, do both.

Several other methods for regenerating α-tocopherol from its radical have been suggested to occur in the body, including recycling by reduced glutathione, GSH. Perhaps the most popular suggestion is recycling by ascorbic acid, vitamin C:

$$\begin{aligned}\alpha\text{-tocopherol}^{\cdot} + \text{ascorbic acid} &\rightarrow \alpha\text{-tocopherol} \\ &+ \text{ascorbic acid}^{\cdot}\end{aligned} \quad (17)$$

Since ascorbic acid is a water-soluble vitamin, which cannot enter the hydrophobic interior of membranes, this mechanism presupposes that the tocopherol˙ radical can move close to the membrane surface for reduction by ascorbic acid outside the membrane. There is much laboratory evidence consistent with reaction (17), but it has not been proved rigorously to operate inside the body. The authors *believe* that it does occur.

Reaction (17) generates an ascorbic acid radical, a fairly unreactive species. Two such radicals, if they meet, can interact to regenerate some ascorbate

$$2 \text{ ascorbate}^˙ \rightarrow \text{ascorbate} + \text{dehydroascorbate} \qquad (18)$$

Several cells contain enzymes that can reduce dehydroascorbate back to ascorbate (using either NADH or GSH as sources of reducing power). Any dehydroascorbate that does not enter cells for regeneration might break down, since it is a very unstable molecule. Hence, ascorbate tends to be lost irreversibly at sites of oxidative damage.

Tocopherols are probably the major chain-breaking antioxidants in human membranes, but not the only ones. Membranes do not contain large amounts of tocopherols: a ratio of one tocopherol per thousand polyunsaturated fatty acid side chains is common.

Repair and replacement

The action of chain-breaking antioxidants leaves some lipid hydroperoxide in the membrane (equations 14 and 15). How can this be removed? One mechanism is *repair*. Some scientists believe that *phospholipase* enzymes remove peroxidized fatty acid side chains from the membrane lipids: they are then replaced by normal fatty acids. The released fatty acid hydroperoxides might then be metabolized by glutathione peroxidase: this enzyme not only acts upon H_2O_2 but also upon free fatty acid hydroperoxides, reducing them to fatty acid alcohols ($-OOH$ to $-OH$). However, the peroxidized fatty acids have to be split from the membrane first. A *phospholipid hydroperoxide glutathione peroxidase* has recently been described in some mammalian tissues. It is apparently capable of acting upon peroxidized fatty acid side chains *within membranes*, but its metabolic importance is not yet clear. The final mechanism for getting rid of peroxidized lipids is *turnover*—membrane constituents in most cells are constantly removed and replaced in the dynamic metabolism of the human body.

Extracellular antioxidant protection

As in the case of cells, the extracellular fluids of the human body (blood plasma, lymph, the fluids lining the lungs and the passages leading to them, seminal plasma, the fluid surrounding the brain and spinal cord (*cerebrospinal fluid*), and the *synovial fluid* (which lubricates joints) have two components that need antioxidant protection. First is the aqueous phase: it is necessary to protect not only essential molecules dissolved in the aqueous phase, but also the surface of cells against oxygen-derived species generated in the aqueous phase. Second is the lipid phase. Protection of this phase is most important in blood plasma and in the fluid lining the alveoli of the lungs. The latter fluid contains *surfactant*—a protein–lipid complex essential in maintaining normal lung function by preventing the alveoli from collapsing.

Blood lipids: targets for free radical attack

Blood plasma contains numerous lipid-containing particles, since it is the vehicle by which lipids are transported around the body (Table 2). Thus *chylomicrons* carry lipids from the intestine to the tissues after a meal. *Very-low-density lipoproteins* (VLDLs) carry fat made in the liver (e.g. from sugars or alcohol—hence the origin of 'sweet tooth obesity' and 'beer belly') to the adipose tissue. *Low-density lipoproteins* (LDLs) deliver cholesterol made in the liver to the tissues and *high-density lipoproteins* (HDLs) may do the opposite (reverse cholesterol transport). All these lipid particles are *lipoproteins*—they contain both lipids and proteins. The lipid portion can readily undergo peroxidation, especially if a diet rich in polyunsaturated fats is eaten, which will eventually raise the content of polyunsaturates in the plasma lipoproteins. Most attention has been paid to

Table 2. *Lipoproteins in human plasma. Each class of lipoprotein contains a range of particles with different sizes and compositions, so the percentages given are only approximations. The structure is similar in each case, with a core of triglyceride and cholesterol ester surrounded by a coat of protein and phospholipid*

Name	Composition (%)					Transport role
	Protein	Phospholipid	Triglyceride	Cholesterol		
				Not esterified	Ester form	
Chylomicrons (CMs)	5	10	65	5	15	Triglycerides and cholesterol from gut to adipose tissue
Very-low-density lipoproteins (VLDLs)	10	15	55	5	15	Triglycerides made in liver to adipose tissue
Low-density lipo-proteins (LDLs)	20	22	12	10	36	Made from VLDL after triglyceride has been taken up by tissues. Supply of cholesterol to tissues
High-density lipoproteins (HDLs)	50	24	4	2	20	Formed in the blood. May remove cholesterol from some tissues

peroxidation of LDLs, because this is thought to be related to atherosclerosis and heart disease (see Chapter 7). Figure 5 summarizes the chemical changes in isolated human LDL after it has been exposed to pro-oxidant conditions. Several molecules disappear before lipid peroxidation accelerates, including tocopherols, β-carotene, and lycopene. The last two are pigments of plant origin: both can, in isolation, react with some free radicals. Whether they actually are important antioxidants in the human body is uncertain as yet.

Although the results in Figure 5 suggest that all kinds of molecules may be important antioxidants in human LDL, the data in Table 3 should help to put them in perspective. The content of these suggested antioxidants is presented in terms of molecules per LDL particle. Each human LDL particle contains, on average, 7 tocopherol molecules. By contrast, most other suggested antioxidants are present in ratios of less than 1 per LDL particle. The most striking example is ubiquinol, which may be an important antioxidant in mitochondria but is present in human LDL only at about 0.5 or less molecules per LDL particle. Putting it another way, one needs to take two or more LDL particles to find one that has even a single ubiquinol molecule. It is, therefore, unlikely that antioxidants other than tocopherol are a major protection of LDLs *in vivo*, since only a

Table 3. *Polyunsaturated fatty acids and antioxidants in human low-density lipoproteins (data courtesy of Professor H. Esterbauer)*

Molecule	Average number of molecules per LDL particle
Total cholesterol	2100
Linoleic acid side chains	1101
Arachidonic acid side chains	153
d-α-Tocopherol	6.4
γ-Tocopherol	0.5
β-Carotene	0.3
Lycopene	0.2
Ubiquinol	< 0.5

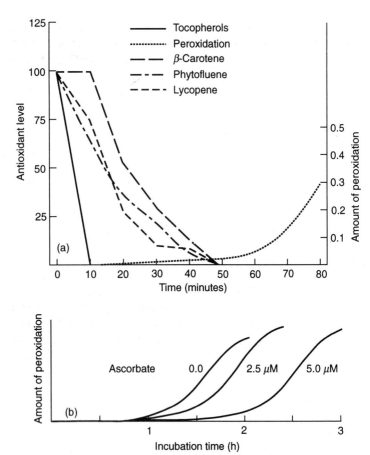

Figure 5. Antioxidant depletion during the peroxidation of low density lipoproteins (LDL) (a) LDLs isolated from human plasma were subjected to oxidative damage (exposure to free copper ions). Not until the antioxidants present had been consumed did LDL peroxidation accelerate. (b) Effect of ascorbic acid on LDL peroxidation. A repeat of experiment (a), but including ascorbic acid in the incubation medium. Note the delay in onset of peroxidation with increasing ascorbic acid concentrations. (Both figures adapted with permission from H. Esterbauer *et al.* (1989). *Free Radical Research Communications*, **6**, 67–75, with thanks to Harwood Academic Publishers.)

fraction of the LDL particles happen to contain them. **The cooperation of α-tocopherol and vitamin C (Figure 5) is probably the most important antioxidant mechanism protecting LDLs in human plasma against peroxidation.**

The aqueous phase

Vitamin C: antioxidant or pro-oxidant?

The aqueous phase of plasma contains a wide range of molecules with antioxidant properties. One of these is ascorbic acid, a very good scavenger of several radicals, as summarized in Table 4. Ascorbic acid may also recycle α-tocopherol in lipoproteins: Figure 5 shows how ascorbic acid delays the onset of peroxidation in isolated human LDLs. If the LDLs are first depleted of α-tocopherol, ascorbic acid has no effect, confirming that it is not protecting against peroxidation directly, but by interacting with α-tocopherol.

In test-tube experiments, ascorbic acid can be made to exert *pro-oxidant* properties (i.e. it increases oxidative damage). Chemically, ascorbate is a reducing agent, able to reduce iron and copper ions. Indeed, it facilitates iron absorption in the gut by this mechanism (see Chapter 2)

$$Fe^{3+} + ascorbate \rightarrow Fe^{2+} + ascorbate^{\bullet} \qquad (19)$$

$$Cu^{2+} + ascorbate \rightarrow Cu^{+} + ascorbate^{\bullet} \qquad (20)$$

Table 4. *Antioxidant properties of ascorbic acid (vitamin C)*

Scavenges superoxide ($O_2^{\bullet-}$) radical
Can scavenge thiyl (sulphur centred) radicals
May reduce some food carcinogens to inactive products
Scavenges singlet O_2 and hydroxyl radical
May regenerate α-tocopherol from α-tocopheryl radicals in membranes
May protect lung lining fluids against damage by air pollutants (such as ozone and oxides of nitrogen) as well as by cigarette smoke

If H_2O_2 is also present, ascorbate can accelerate $OH^{\cdot}$ formation by forming Fe^{2+} and Cu^+, which react with H_2O_2 to give $OH^{\cdot}$. Mixtures of iron salts and ascorbic acid are frequently used to accelerate lipid peroxidation in isolated membranes.

Why do these pro-oxidant effects of ascorbic acid not usually happen *in vivo*? **Simply because, under most circumstances, free iron and copper are not available in the extracellular fluids.** How is this achieved?

In plasma from healthy humans, the iron transport protein transferrin is normally only 20–30 per cent loaded with iron and keeps the concentration of free iron in plasma at effectively zero (see Chapter 2). Iron bound to transferrin will not participate in radical reactions, and the available iron-binding capacity in plasma gives this fluid a powerful antioxidant property toward iron-stimulated radical reactions. Similar considerations apply to the protein *lactoferrin*, which, like transferrin, can bind two molecules of Fe^{3+} per molecule of protein, and prevent the iron from undergoing free radical reactions. Human tear fluid and nasal secretions contain lactoferrin—presumably it acts both to hinder free radical damage and to decrease the growth of bacteria, by depriving them of the iron that they need.

Copper is also safely bound in plasma. The major copper-containing protein in plasma is ceruloplasmin, unique for its intense azure-blue coloration. Caeruloplasmin has *antioxidant* properties: it can inhibit peroxidation of most lipids and it oxidizes Fe^{2+} to Fe^{3+} without formation of oxygen radicals. There is some suggestion that one of the copper ions on caeruloplasmin can oxidize LDL, however. Any plasma copper not attached to ceruloplasmin is probably bound to albumin; again this may be a defence mechanism because any radicals produced by reaction of the copper with peroxides will react with the albumin and not be allowed to escape into 'free solution' to damage more important targets. Albumin provides thiol (−SH) groups at a concentration of 0.3–0.5 mM in human plasma—these are capable of scavenging a wide range of radicals. Indeed, the authors have proposed that, among its many other biological functions (which include transport of fatty acids and binding of toxins), albumin is a 'sacrificial antioxidant'—it can be damaged by a wide range of oxidants, and the damaged albumin is then simply removed from the circulation and replaced.

Enzymes: antioxidants in extracellular fluids

What seems to be of limited importance in human plasma are the three antioxidant defence enzymes, SOD, catalase, and glutathione peroxidase. Human plasma contains a low activity of glutathione peroxidase, but there is almost no GSH to allow it to function. No catalase is present. SOD levels are very low—there are traces of Cu,Zn-SOD, Mn-SOD, and of another type, EC-SOD (extracellular SOD). EC-SOD may normally exist bound to the surfaces of vascular endothelial cells surfaces, from where it sometimes falls off to leave a low activity in the plasma. It should also be noted that tissue injury can release antioxidant enzymes from broken cells, so raising their levels in extracellular fluids and helping to diminish the general pro-oxidant actions of cell disruption (discussed in detail in Chapter 6).

Why should there be so little of the cellular antioxidant defence enzymes in normal extracellular fluids? By allowing the survival of $O_2^{\cdot-}$ and H_2O_2 in the extracellular environment for a short period, the body may sometimes utilize these molecules as messengers. Thus, at sites of inflammation, $O_2^{\cdot-}$ and H_2O_2 produced by activated white blood cells might influence the behaviour of the other cell types present. It is *essential* to keep transition metal ions out of the way, so that dangerous species such as $OH^{\cdot}$ are not made. Thus, in the authors' view, the primary antioxidant defence in human plasma is the safe binding or inactivation of metal ions—this allows use of $O_2^{\cdot-}$ and H_2O_2 as signal molecules, and also allows vitamin C to exert its specialized antioxidant properties. After performing its useful functions, H_2O_2 (which crosses cell membranes readily) can diffuse into the nearest cell for metabolism by intracellular catalase or glutathione peroxidase. Platelets, white blood cells, endothelial cells, and red blood cells could all act as 'sinks' for H_2O_2 in this way.

Superoxide may disappear by non-enzymic dismutation to H_2O_2:

$$2O_2^{\cdot-} + 2H^+ \rightarrow H_2O_2 + O_2 \tag{21}$$

Red blood cells also have a channel in their membranes through which $O_2^{\cdot-}$ can pass, to be removed by intracellular Cu,Zn-SOD in these cells (red blood cells have no mitochondria, so their only SOD is the CuZn type).

Uric acid

Uric acid, present in human body fluids at concentrations in the range 0.3–0.5 mM, has also been suggested to exert antioxidant properties but the extent to which this happens is uncertain. In the fluid lining the upper respiratory tract in humans, uric acid may be an important scavenger of inhaled oxidizing air pollutants, such as ozone and nitrogen dioxide.

4 Antioxidant vitamins and nutrients

Elixirs of youth or tonics for tired sheep?

T.L. Dormandy (1977)

Antioxidant defences rely heavily on vitamins and minerals from the diet (see Table 1). Indeed, research in Jamaica has shown that the problems suffered by children with *kwashiorkor*, a disease caused by lack of suitable dietary protein, arise not only from the deficiency in protein itself, but also from inadequate levels of antioxidants (including caeruloplasmin, tocopherols, and GSH) and from failure to make enough albumin, or sufficient transferrin to ensure safe binding of iron in the plasma.

Would people in 'advanced' countries gain any benefit from consuming 'supplements' of these nutrients, in addition to those absorbed from the diet? The companies that manufacture 'vitamin E', 'vitamin C', 'β-carotene', 'selenium', 'multivitamin tablets', and SOD capsules (with or without added catalase) would like us to think so. This is a multi-million dollar business in the USA and is growing to a similar size in other parts of the world. Some claims are unlikely to be true. SOD and catalase are proteins—in the gut, they will be digested by enzymes into amino acids and peptides. Little, if any, will be absorbed intact. No more will be absorbed from SOD/catalase tablets than from uncooked foodstuffs (most of which contain these enzymes). Some of the other claims deserve careful attention, however. Several of them are analysed in detail later in this chapter, and Chapter 7 is specifically devoted to a consideration

Table 1. *Nutrients and antioxidant defence*

Nutrient	Examples of useful roles in the human body
Iron	Catalase, correct functioning of mitochondria, haemoglobin
Manganese	Mn-SOD in mitochondria
Copper	Cu,Zn-SOD, caeruloplasmin
Zinc	Cu,Zn-SOD: more generalized antioxidant properties? Stabilizer of membrane structure?
Proteins	Sulphur-containing amino acids are needed to make GSH. SODs, catalase, glutathione reductase and peroxidases, metal transport, and storage proteins. Albumin may be a 'sacrificial' antioxidant carrier of copper in plasma (Chapter 2)
Riboflavin (water-soluble B-vitamin)	Glutathione reductase, correct functioning of mitochondria, needed to make FMN and FAD
Selenium	Glutathione peroxidases; thyroid function; may aid detoxification of carcinogens?
Vitamin E (tocopherols; fat-soluble vitamin)	Protection against lipid peroxidation; may also help to stabilize membrane structure
Vitamin C (ascorbic acid; water-soluble vitamin)	Hydroxylase enzymes, water-soluble antioxidant, recycles vitamin E (?), reduces nitrosamine carcinogens.
β-carotene	Precursor of vitamin A. *May* have some antioxidant properties—powerful scavenger of singlet O_2, may react with peroxyl radicals. Some reports that it inhibits lipid peroxidation in membranes, but only at low O_2 concentrations

Nutrient	Examples of useful roles in the human body
Lycopene	Orange-red pigment in tomatoes. Powerful scavenger of singlet O_2. *Suggested* to be an antioxidant *in vivo*, but this has not yet been established.
Retinol (vitamin A; fat-soluble vitamin)	Some antioxidant properties demonstrated *in vitro*, but no good evidence that it acts as an antioxidant *in vivo*
Nicotinamide (a B-vitamin)	Needed to make NAD^+, NADH, $NADP^+$, NADPH—needed for glutathione reductase. Important in cell metabolism and energy production

of nutrition, free radicals, and cardiovascular disease. As a preliminary, let us see what is known about dietary needs for antioxidant vitamins and minerals.

Vitamin E (tocopherols)

Tocopherols are prone to undergo oxidation, especially if exposed to light and air. To limit this, they are often sold as more stable esters, vitamin E acetate, or vitamin E succinate (see Chapter 3, Figure 4). After ingesting vitamin E esters, enzymes release 'free' vitamin E for absorption (in the presence of bile) in the duodenum and small intestine. Vitamin E is fat-soluble; you need dietary fat to absorb it. Once absorbed, vitamin E is transported in the blood in the lipid component of plasma lipoproteins, such as low-density lipoproteins. Laboratory studies have established that vitamin E is an extremely effective chain-breaking antioxidant that protects polyunsaturated lipids from free radical damage (see Chapter 3). Vitamin E appears to be essential for the protection of circulating lipoproteins and the correct functioning of cell membranes. Its possible role in protecting against cardiovascular disease is considered in detail in Chapter 7.

Do humans need vitamin E?

What is the evidence that humans need vitamin E in their diet? The clearest evidence is provided by diseases in which tocopherol absorption from the gut is halted. Because tocopherols are fat-soluble, they enter the body dissolved in fats. Several diseases are known in which impaired fat absorption eventually leads to severe depletion of tocopherols. The best example is *abetalipoproteinaemia*, an inborn disease in which dietary fat is digested and absorbed, but not transported out of the gut. Degeneration of the nervous system resulting from this disease can largely be prevented by large oral doses of vitamin E (to ensure that at least some is absorbed). Newborn babies have low concentrations of vitamin E in their plasma, especially premature babies. This may predispose their red blood cells to rupture ('haemolytic anaemia of prematurity') and their retinas to degenerate (sometimes causing blindness) when excess O_2 is administered (it is often necessary to increase the O_2 content in incubators housing premature babies to ensure that enough O_2 enters the blood). The incidence of the haemolytic anaemia of prematurity and the severity of 'retinopathy of pre-maturity' have decreased with the introduction of vitamin E supplementation of premature babies.

But *how much* vitamin E do humans actually need? Although recommended dietary allowances (RDAs) have been set (Table 2), the question is in fact difficult to answer, because short-term deprivation of vitamin E in humans does no obvious harm. The RDAs may be just enough, an excess, or suboptimal, in the long term. In animals, however, symptoms of vitamin E deficiency are well described: they range from death of the fetus in pregnant female rats (the original test by which vitamin E was discovered: see Chapter 3) to 'white muscle disease' in lambs and calves and brain degeneration in chicks. Hence veterinary surgeons have laid down appropriate dietary vitamin E requirements for animal species. These requirements are affected by the selenium content of the diet (a high selenium intake can 'spare' vitamin E to some extent, meaning less vitamin E is required) and by the lipid composition of

Table 2. *Recommended dietary allowances (RDAs)[†] or Reference Nutrient Intakes (RNI) for various nutrients (values are quantities needed per day to meet the known nutritional needs of healthy persons)*

	UK		USA	
	Males	Females	Males	Females
Vitamin A	700 μg	600 μg	1000 μg	800 μg
Vitamin E*	>4 mg	>3 mg	10 mg (15 IU)	8 mg
Vitamin C[†]	40 mg	40 mg	60 mg	60 mg
β-Carotene	not set	not set	not set	not set
Selenium	75 μg	60 μg	50–200 μg	
Iron	8.7 mg	14.8 mg	10 mg	
Zinc	9.5 mg	7.0 mg	15 mg	
Copper	1.2 mg	1.2 mg	2–3 mg	
Fibre[†]	12–14 g		20–35 g	

* 1.49 International Units (IU) of vitamin E are equivalent to 1 mg of d-α-tocopherol. 1 mg of synthetic dl-α-tocopherol acetate (all-*rac*-tocopherol) equals 1.0 IU. The intake recommended is expressed as mg of d-α-tocopherol (RRR-α-tocopherol) equivalents. Multiplication factors are used to take into account the lower biological effectiveness of synthetic tocopherols (Chapter 3, Figure 4). Before taking these calculations too seriously, remember that 'biological effectiveness' of vitamin E is based on a rat assay which is not necessarily relevant to humans.
[†] The term 'dietary fibre' lacks a precise definition. The UK RDA refers to total 'non-starch carbohydrate polymers'.

the diet—the more unsaturated fats that are eaten, the more vitamin E is required. Probably this is also true of humans, especially in view of proposals that we should increase the unsaturated–saturated ratio of our dietary lipids (see Chapter 7).

The peroxidation of low-density lipoproteins (LDLs) is thought to be important in cardiovascular disease. Figure 1 shows an experiment in which LDLs were isolated from the plasma of human

[†] Smokers probably need a higher intake by up to 80 mg/day.
[†] RDA can also mean Recommended daily amounts.

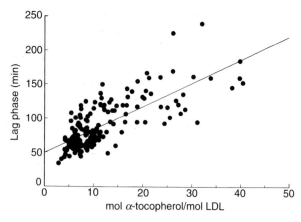

Figure 1. Tocopherols and the peroxidizability of human low-density lipoproteins (LDL): the relationship between the oxidation resistance of LDL and its α-tocopherol content. The peroxidation resistance of LDL isolated from human volunteers is not significantly correlated to the content of vitamin E (usually 3–15 molecules α-tocopherol per LDL particle). However, if subjects consuming extra oral vitamin E, so raising the vitamin E content of LDL are included, a correlation is seen. Peroxidation resistance is measured as the length of the 'lag phase' before peroxidation accelerates after exposure to copper ions. Data are also included from isolated LDL that has been loaded with vitamin E by incubation in the test tube before analysis. (From Esterbauer *et al.* (1991). *Annals of Medicine*, **23**, 580. Reproduced with permission.)

volunteers and their 'peroxidizability' determined. The lag period before the onset of peroxidation was taken as an index of the 'antioxidant content' of the LDLs (see Chapter 3, Figure 5). Surprisingly, the length of the lag period did not correlate with the LDL content of vitamin E. However, if the subjects first consumed vitamin E supplements to raise the amount of vitamin E in the LDL, the lag periods increased and a correlation with vitamin E content was observed. It is difficult to raise the vitamin E content more than two- or threefold, and within about a week of ceasing dietary supplementation the LDL vitamin E level returns to normal. The data also indicate the amounts of vitamin E that need to be

Table 3. *Effect of dietary vitamin E on the tocopherol content of human LDL and on its resistance to peroxidation. Human volunteers were given daily oral doses of RRR-α-tocopherol for 3 weeks. Results are presented as average percentages. Six days after the supplements were stopped, the oxidation resistance of the LDL had returned to normal (data abstracted from Esterbauer, H.* et al. *(1991).* Annals of Medicine, **23**, *573–81*

Daily Supplementary Dose (International units)	α-Tocopherol in LDL (% of control)	Oxidation resistance of LDL, lag phase (% of control)
0 (control)	100	100
150	138	118
225	158	156
800	144	135
1200	215	175

consumed to increase the antioxidant protection of LDLs (Table 3). However, we do not necessarily *know* that this is worth doing, a point considered further in Chapter 7.

The richest natural sources of vitamin E are vegetable oils (including salad oils and margarines), nuts, and whole grains. Wheat germ oil is the single richest source. Vitamin E in foods is slowly oxidized, and this is accelerated by heating and light. Vitamin E is not completely absorbed in the intestine, and at least half of any dietary supplements probably end up 'down the pan', especially if dl-α-tocopherol rather than d-α-tocopherol is consumed.

β-Carotene and vitamin A

Carotenoids are a class of molecules, widely distributed in nature, that give plants, fruits, and vegetables many of their bright colours. They are not essential for human life and so no RDA is set for them (Table 2). *β*-Carotene is a major contributor to the pigments of carrots, broccoli, tomatoes, red peppers, pumpkins, and winter squash. Over 600 carotenoids are known to exist in nature, and some 50 of these can serve as precursors of the fat-soluble vitamin known as vitamin A (sometimes called *retinol*). *β*-Carotene is often referred to as *pro-vitamin A* for this reason. Failure to maintain an adequate intake of carotene-rich fruits and vegetables can result in low vitamin A levels, a common nutritional deficiency in many third-world countries. Among other roles, vitamin A is needed for the retina of the eye to function properly. Simple lack of this vitamin and of the carotenoids that give rise to it accounts for some 100 000 cases of infantile blindness worldwide.

Epidemiological[†] studies have suggested that β-carotene is an important dietary anticarcinogen (cancer preventing agent) as well as a protective factor against heart disease (see Chapter 7). Diets rich in yellow and green vegetables seem to be protective against cancer (particularly lung cancer and cancer of the cervix) and heart disease, since there is a correlation between body carotenoid content (a reflection of a vegetable-rich diet) and lower incidence of these diseases. However, this does not *prove* that the carotenoids are responsible, rather than any other agent (or combination of agents) present in these foods. To put it simply, **correlation does not imply causation**, a principle that must always be born in mind when interpreting epidemiological studies (see the Appendix to this chapter). An amusing example of this was published in the scientific journal *Nature*: the decline in the birthrate in Germany is highly correlated with the decline in the number of breeding storks.

Vitamin A, like β-carotene, has also been suggested to be a dietary anticarcinogen from studies of intake versus cancer incidence. Dietary vitamin A comes mainly from foods of animal origin, whereas β-carotene comes from plant sources. High nutritional supplements of vitamin A are not without health risk, and have been reported to cause headaches, drowsiness, insomnia, irritability, joint and bone pain, loss of hair, dry skin, and fatigue. If it is thought that more vitamin A is needed, it would seem sensible to supplement the diet (if indeed such supplementation is ever required) with β-carotene, which will be converted to vitamin A as required by the body.

β-Carotene and other carotenoids are important antioxidants in

---

[†] For a brief commentary on epidemiology, see the Appendix to this chapter.

plants, helping to scavenge singlet O_2 that can be generated by the interaction of light with the plant pigment chlorophyll (see Chapter 1), before the singlet O_2 can damage the plant. Some evidence from test-tube experiments has suggested that β-carotene might exert other antioxidant effects, for example by inhibiting lipid peroxidation (Table 1). However, there is no evidence as yet for an antioxidant action of carotenoids in the human body. Even if one interprets the epidemiological evidence as indicating that β-carotene does have an anti-cancer role, this does not mean that it is related to an antioxidant activity of this molecule rather than to other effects it might exert, such as increasing the levels of vitamin A. Whereas loading LDLs with vitamin E increases their resistance to oxidative damage (Figure 1), loading them with β-carotene does not.

Vitamin C (ascorbic acid)

Ascorbic acid is synthesized by most animals except the guinea pig, primates (including humans), the fruit bat, and some birds. Our essential requirement for ascorbic acid is therefore met entirely by dietary intake of this water-soluble vitamin. Consequences of ascorbic acid deficiency were recognized long before its nutritional functions were known. The Egyptians, Greeks, and Romans, and the Spanish, Portuguese, and English sailors, were aware of the lethal effects of the disease scurvy. The British Royal Naval physician Lind is credited with curing scurvy in 1753 by treating affected sailors with oranges and lemons (formerly called limes— hence the nickname 'limeys' for British sailors).

Ascorbic acid was first isolated from plant and animal tissues by Szent-Györgyi in 1928 and first chemically synthesized in 1933. Good sources of ascorbic acid are turnip greens, green peppers, broccoli, brussel sprouts, paprika, cauliflowers, cabbage, tomatoes, new potatoes, oranges, lemons, and other citrus fruits.

How much vitamin C do humans need?

The normal concentration of ascorbic acid in human blood plasma ranges from 0.8 to 1.6 mg per 100 ml and a concentration of 0.2 mg per 100 ml or less is said to be indicative of 'deficiency'. Ascorbic acid concentrations within cells are often higher. The RDA (Table 2) is set at 40 mg in the UK and 60 mg in the USA; both levels are more than adequate to prevent scurvy. After ingestion of gram quantities of vitamin C, plasma levels may transiently reach 2.0–2.5 mg per 100 ml. It is difficult to achieve higher levels because of efficient excretion by the kidney. Despite this, some authors have recommended intake of gram quantities of vitamin C every day. Is this safe? Frequent oral intake of large amounts of vitamin C may produce gastric irritation with diarrhoea and acidification of the urine, which has been said to predispose to kidney stone formation. One end-product of dehydroascorbate breakdown is oxalic acid, which has also been suggested to encourage stone formation. However, the authors are not aware of convincing *clinical* evidence that mega-dose ascorbic acid does increase the incidence of kidney stones. Claims that large daily mega-doses of vitamin C help patients with cancer to live longer, or that they cure the 'common cold', have not been supported by well-designed clinical trials. What vitamin C *might* do is to chemically reduce certain carcinogens in food, such as nitrosamines, to harmless species. Thus adequate vitamin C intake might help to prevent certain types of cancer, rather than having any great use in the treatment of established cancer. What amount is 'adequate' in this context is unknown.

Several epidemiological studies have found a correlation between low intakes (or low blood levels) of vitamin C and increased risk of cancers, especially of the oesophagus, mouth, pancreas, and stomach. As with all epidemiological studies of this type, one cannot draw the conclusion that vitamin C protects against cancer. High intakes or blood levels of vitamin C are a 'marker' of a diet rich in fruits and vegetables, and it could be any factor, or combination of factors (including carotenoids!) in such a diet that is protective (Table 4).

Ascorbic acid levels in the fluids that line the alveoli of the lung are higher than in plasma, and may help to protect the lung against certain gaseous air pollutants, such as ozone (O_3), the free radical nitrogen dioxide ($NO_2^\bullet$), and cigarette smoke. Seminal fluid also has

Table 4. *Potential cancer fighters in plants (other than vitamins E and C and carotenoids)*

Compound	Possible action	Food sources
Organic sulphides	Stimulation of enzymes that detoxify carcinogens	Garlic, onions
Catechins	Antioxidants? Directly cytotoxic to cancer cells??	Green tea, black tea, many berries
Flavonoids	Antioxidants? Directly cytotoxic to cancer cells?? Prevention of binding of hormones needed for cancer growth	Most fruits and vegetables
Phytic acid	Binds metals, decreases iron absorption	Many grains
Genistein	May block growth of new blood vessels into growing tumours	Soybeans
Limonoids	Induce protective enzymes	Citrus fruits
Fibre	Increases speed of movement of faeces through colon, dilutes carcinogenic compounds and delays their formation	Grains, vegetables
Isothio-cyanates	Induce protective enzymes	Mustard, radishes

higher ascorbate levels than plasma, and there are suggestions that ascorbate is needed for normal sperm function.

Pro-oxidant properties of vitamin C?

Ascorbic acid has many antioxidant properties (see Chapter 3, Table 4), but it can also reduce iron and copper ions and accelerate oxidative damage. The fact that high doses of ascorbic acid can be consumed regularly, and apparently safely, by healthy adults indicates how well the availability of these metals to catalyse free radical reactions is normally controlled *in vivo* (see Chapter 3).

However, in disease states, metal ions may become more 'available' (see Chapter 6) and so one could speculate that ascorbic acid *might* make things worse. Thus giving supplementary ascorbic acid to patients suffering from iron overload diseases such as idiopathic haemochromatosis (see Chapter 2) can worsen tissue injury, unless the patients are simultaneously receiving an iron ion chelating agent that binds the free iron and prevents it from catalysing free radical reactions.

Selenium

Selenium has been recognized as an essential trace element for animals and humans for over 30 years, although plants do not seem to require it. RDA values are 50–200 μg per day (Table 2). Epidemiological studies suggest that humans with a low selenium intake are at a greater risk of developing cancer and cardiovascular diseases.

Animal studies show that selenium is protective against cancers which are caused by the active metabolism of certain chemicals into cancer-causing agents by the liver: selenium seems to induce the synthesis of enzymes that 'detoxify' the carcinogens. Selenium deficiency can occur in people living in areas of the world with low selenium levels in the soil, such as certain areas of China, New Zealand, or Finland. Selenium deficiency in Finland has been corrected since 1984 by deliberately introducing a selenium compound into the environment by adding specially enriched fertilizers to the soil. The most striking consequences of human selenium deficiency have been observed in the People's Republic of China. *Keshan disease* is a degenerative disease, particularly affecting the heart, that is prevented by supplementation of the diet with a selenium salt. The name comes from an episode in 1935 in which 57 out of 286 inhabitants of a village in Keshan County of Heilongjiang Province died of the disease. Careful epidemiological studies showed that the incidence of the disease was correlated with that of various animal diseases known to be related to selenium deficiency, such as white muscle disease. Selenium deficiency has also been implicated in *Kashin–Beck disease*, a disabling joint disease seen in children in areas of North China, North Korea, and East Siberia. The name comes from the Russian scientists who first described the disease.

Although selenium is frequently described as an antioxidant in advertisements for nutritional supplements, its only established antioxidant role lies in its function at the active sites of glutathione peroxidase enzymes (i.e. selenium is not *itself* an antioxidant). Only small amounts of dietary selenium are needed to gain maximal glutathione peroxidase activity (50 µg per day), less than the average dietary intake in most countries. In excess (>350–600 µg per day), selenium is toxic. An early sign of excess selenium intake is deformation and loss of fingernails and toenails. The authors feel that, except in a few special geographical areas, dietary selenium supplements have not been shown to be necessary.

Iron: boon or bane?

In severe iron deficiency, not enough haemoglobin can be made and transport of oxygen by the blood is impaired. In fact, iron deficiency anaemia to an extent that affects health is quite rare in Western Europe and the USA since the RDA (Table 2) is easily met. Subjects at risk include poorly nourished pregnant women and growing children. **In children, an adequate intake of iron is *essential* for mental development, and great care should be taken to avoid deficiency.** Severe iron deficiency is associated with increased susceptibility to infections.

However, how much iron do adults need? In Europe and the USA, there is widespread fortification of foodstuffs with iron salts or metallic iron. This fortification can cause problems for food manufacturers, because adding iron salts to foods can stimulate oxidative damage to lipids (lipid peroxidation—see Chapter 3) and to other food constituents, especially if the iron is in a form that can be reduced by ascorbic acid or other reducing agents in the food to give Fe^{2+}. **In the USA, it has been estimated that the incidence of real iron-deficiency anaemia is lower than that of undiagnosed iron-overload disease, so the alleged benefits of widespread iron supplementation need much more clinical evaluation than they have been given to date (Table 2 in Chapter 2 lists the devastating**

consequences of prolonged iron overload in haemochromatosis patients). Several suggestions that high body iron stores in men correlate with increased incidence of heart disease and cancer have been made in the medical literature. Tissue iron stores appear to increase with age in men, and after the cessation of menstruation in women (Figure 2). Presumably this iron resides largely within the storage protein ferritin.

In third-world countries, the body's protective response to chronic infections with bacteria and other parasites includes low blood iron level, often mistaken for iron-deficiency anaemia.

Zinc and copper

Zinc is essential for the function of several enzymes in the human body, including Cu,Zn-SOD (Table 1). Because zinc has a fixed valency of 2 (it forms only Zn^{2+} ions), zinc salts (unlike iron and copper salts) cannot promote free radical damage. For example, zinc salts do not form hydroxyl radicals ($OH^{\cdot}$) from H_2O_2. Several laboratory experiments have suggested an antioxidant action for zinc because it can bind to biological molecules to block the binding

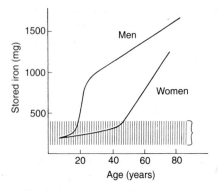

Figure 2. Increased body iron stores with age in US males and females. Note that women's body iron stores stay low until after the menopause. Perhaps the shaded range is therefore the optimal range of body iron stores. (Reproduced with permission from Professor Randall Lauffer.)

of iron and copper and so protect them against oxidative damage mediated by these two metals. Whether these effects actually occur in the human body is uncertain.

Zinc also appears to be necessary for proper functioning of the immune system, and it is thought to 'stabilize' membranes in some undefined way. The RDA for zinc (Table 2) is easily met in the UK and USA and so zinc deficiency is very rare. Consuming large amounts of zinc salts can decrease absorption of copper from the gut: indeed, this is one recommended therapy in patients with 'copper overload' diseases. However, it may not be a good idea in normal subjects, since copper is essential for many biological functions, including the synthesis of the copper-containing proteins Cu,Zn-SOD, caeruloplasmin, and cytochrome oxidase (Table 1). Copper also appears to be necessary in some way to make haemoglobin. Copper deficiency in humans is rare; it is more of a problem in veterinary medicine, since several animal species (e.g. cattle) seem predisposed to develop copper deficiency. There are suggestions that premature babies may be at risk of copper deficiency. However, in view of the powerful pro-oxidant actions of copper (see Chapter 2), the authors do not believe that dietary supplementation is needed. Indeed, in a recent study of Finnish men, high copper concentrations in blood plasma were correlated with an increased incidence of heart attacks (remember the limitations in interpreting epidemiological correlations, however).

Fish oils

Rancidity of fats and oils has been recognized as a problem since antiquity: it is caused by the chain reaction of lipid peroxidation (see Chapter 3, Figure 3). Polyunsaturated fats are much more prone to this process than saturated or monounsaturated fats. Our consideration of vitamin E led us to suggest that increased intake of PUFAs might require increased intake of tocopherols, although many foods with high PUFA levels (such as vegetable oils) are themselves good sources of vitamin E, provided that they have not been stored too long or mishandled.

So, is a high dietary intake of polyunsaturated fats bad for us? One area in which this question has been asked is in relation to cardiovascular disease, discussed further in Chapter 7. Another is that of the use of (n−3) fatty acids in clinical medicine. Interest in these fatty acids resulted from the pioneering work of Dyerberg with Greenland Eskimos.

The (n−3) fatty acids have their first double bond at the third carbon from the methyl end of the chain. The simplest is α-linolenic acid: others are eicosapentaenoic acid and docosahexaenoic acid (see Chapter 1, Table A1 and Figure 3). α-Linolenic acid is found in some vegetable oils (see Chapter 1, Table A1) whereas the last two fatty acids are notably enriched in fat from certain fish, particularly herring, salmon, and mackerel. Epidemiological studies among the Greenland Eskimos showed a much lower frequency of heart attacks, diabetes, the skin disease psoriasis, and multiple sclerosis than in UK or USA populations, and it was speculated that their lower intake of saturated fat and higher intake of (n−3) fats might be responsible. Several other epidemiological studies showed that consumption of fish is associated with lower mortality from heart disease, in several different populations.

There have therefore been proposals that consumption of fish oils could be beneficial, and trials of fish oils as therapy for various

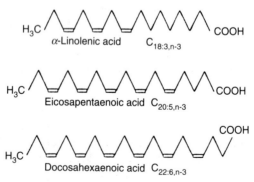

Figure 3. Structures of the main (n–3) fatty acids.

diseases, including arthritis, psoriasis, and inflammatory disorders of the intestine have been carried out, with no really convincing evidence of benefit. One problem is that the (n−3) fatty acids in fish oils peroxidize at a very high rate and therefore have a very short storage life. The authors have found lipid peroxides in 20–30 per cent of commercial fish-oil preparations bought over the counter in drugstores. Thus, in tests of the efficacy of fish oils in medical treatment, it is difficult to be certain what patients are actually getting unless the degree of peroxidation of the administered material is considered. An increased intake of polyunsaturates may increase the vitamin E requirement. Perhaps the best way of taking fish oils is to eat fish, ensuring that it is *fresh* fish.

Food antioxidants

Polyunsaturated fats in the human body are readily attacked by free radicals, forming lipid peroxides (see Chapter 2). Food scientists are also concerned, because the more polyunsaturates in a food, the more likely is peroxidation to occur. Thus, in principle, margarines rich in polyunsaturates have a much shorter shelf-life than most butters, which are richer in saturated and monounsaturated fats. Polyunsaturate-rich cooking oils pose a particular problem. At high temperatures, peroxides decompose to produce a range of unpleasant-tasting and foul-smelling products such as epoxides, aldehydes, acids, and ketones. The 'age' and temperature of oils used in deep frying must therefore be carefully regulated. Foods cooked in peroxidized oils pick up a rancid smell and taste. Similarly, oxidized lipids in mishandled foods impart 'off flavours', such as the 'warmed-over flavour' that can develop in chicken that has been cooked, cooled, and then re-heated.

Food manufacturers can control peroxidation by several mechanisms:

(1) Sealing foodstuffs under nitrogen or in vacuum packs. If there is no oxygen, there can be no peroxidation (until the pack is opened).
(2) Adding chelating agents to bind metal ions that might accelerate

free radical reactions in food. Iron and copper are a particular nuisance. For example, the grinding and processing of meat can release iron from the cells.

(3) Adding chain-breaking antioxidants that stop the chain reaction of lipid peroxidation by reacting with peroxyl radicals. Thus vitamin E (α-tocopherol) is often added to foods: it acts as a chain-breaking antioxidant (see Chapter 3)

$$\alpha-TH + lipid-O_2^{\cdot} \rightarrow \alpha T^{\cdot} + lipid-O_2H \tag{1}$$

Synthetic chain-breaking antioxidants that act in exactly the same way and have been used in food include

- butylated hydroxyanisole (BHA),   E320
- butylated hydroxytoluene (BHT),   E321
- propyl gallate,   E310[†]
- nordihydroguaiaretic acid (NDGA),   –

There have been repeated concerns about the safety of these synthetic antioxidants, often fuelled by experiments in which grotesquely high doses were shown to produce toxicity in small animals. There is therefore a growing interest in 'natural' anti-oxidants, already found in living organisms. Plants contain a wide range of phenols[†] that have chain-breaking antioxidant ability and can extend the shelf-life of polyunsaturated fats. For example, red wine contains phenols that inhibit LDL oxidation in the test tube. The *flavonoids* have probably been the most studied plant phenols. However, flavonoids and other plant phenolics have many other effects. 'Natural' must never be equated to 'safe'—remember that the toxin cyanide is produced naturally by many plants, as are a number of cancer-causing chemicals such as aflatoxin (a powerful inducer of liver cancer, produced by a fungus).

[†] A phenol contains a benzene ring with a hydroxyl (−OH) group attached. Vitamin E is a phenol.

[†] 'E' numbers identify food additives recognized as safe within European Economic Community countries.

Food irradiation

The treatment of foodstuffs with ionizing radiation (such as X-rays or γ-rays) to kill insects and bacteria, or to prevent germination or ripening, is slowly becoming accepted in many countries. Strict laws govern the types of food that may be irradiated and the dose of radiation used. However, enforcement of these laws requires reliable methods that can detect the consequences of radiation exposure. Measurement of end-products of free radical attack upon lipids (lipid peroxides), proteins, or DNA (such as 8-hydroxy-guanine—see Chapter 1, Figure 5) is one approach. Irradiation causes some depletion of antioxidants in food, and the formation of products of free radical damage to the food matrix. Remember that water is split by ionizing radiation to yield hydroxyl radicals (see Chapter 1, Figure 6), which can attack all food molecules and set off chain reactions. Irradiation does not make food radioactive, but many scientists are worried that it could be used to 'clean up' high bacterial contamination of food due to poor hygiene or handling practices.

Appendix to Chapter 4
Clinical epidemiology. What it can and cannot tell us

Statistical techniques have been developed and refined to help detect the causes of human disease. Since these techniques were first applied to infectious diseases, this medical discipline is called *epidemiology*, a term originally referring to its application to epidemic diseases such as smallpox, plague, and so on. Epidemiology attempts to relate the occurrence of disease in a given population to some *common factor*. Early examples include the relation of cholera to consumption of infected drinking water, and the association of some forms of lung cancer with cigarette smoking.

Most epidemiological studies are either *retrospective* or *prospective*. In the former, patients with a particular disease are studied to identify aspects of their lifestyle that may be common to them all. By contrast, prospective studies examine a population and see what happens to the individuals with time. An example would be measuring the level of selenium in the blood of 5000 patients and then following them to see which developed cancer or heart disease.

The main problem with epidemiology is that correlation does not imply causation. Only *intervention* can prove this, and then only after very careful analysis of the results. Thus, suppose that all patients with heart disease show low blood levels of compound X. Perhaps lack of X causes heart disease. Equally likely, however, is the possibility that the real cause of both phenomena is something different. In an *intervention trial* one would supplement a defined group of patients with X and compare their fates with those of a 'control' group. One major task of the epidemiologist is the careful selection of appropriate control groups for comparison with the group of patients being studied. For example, suppose that one group of patients is known to consume large amounts of vitamins E and C and also shows lower incidence of heart disease than another group. Does this mean that vitamins E and C prevent heart disease? Not necessarily, since consumers of vitamin pills might be more interested in their health than average, so that they could be less obese and smoke less. Poor people suffer more disease than richer people, and poor people may not be able to afford vitamins. All of these *confounding factors*, and others, must be taken into account in the analysis of data from patients and controls.

Illustrations of the problems

The health professional studies
In May 1993, two papers were published in the *New England Journal of Medicine* (volume 328, pages 1444–9 and pages 1450–6). The results were greeted with enthusiasm in the lay press (e.g. Figure A1), and there were even requests for changes of the policy of the US Food and Drug Administration. 'FDA's current policy towards health claims restricts consumers' access to useful information needed to make healthy dietary choices. FDA has narrowly interpreted the Nutrition Labeling and Education Act of 1990 (NLEA) to prevent manufacturers from distributing scientific information about the relationship between diet and health. Unless Congress redirects FDA to allow the flow of accurate and truthful scientific information to the consumer, results from studies such as in the *New England Journal* will be confined to a narrow group within the scientific community. Without such legislative direction, FDA will continue its narrow interpretation of NLEA which restricts access to important information consumers want.'[†]

What is it all about?

[†] Reproduced with permission from *Vitamin Issues*, an update on the rationale for supplement legislation, published by the Council for Responsible Nutrition, an association of the nutritional supplements, ingredients, and other nutritional products industry.

*Vitamin E Greatly Reduces Risk
Of Heart Disease, Studies Suggest*

Best Results Found in Those Taking Large Doses   (*New York Times*)

*Supplements of Vitamin E Reduce
 Heart Disease, Two Studies Say*   (*New York Times*)

What's
 So Hot
 About
 Vitamin E?

(*Washington Post*)

Figure A1. Some of the newspaper headlines that greeted the nurses study.

First paper

In 1980, 87 245 female nurses aged 34–59, free of obvious cardio-vascular disease, completed dietary questionnaires that aimed to assess their consumption of a wide range of antioxidants, including vitamin E. During follow-up of up to 8 years, 437 non-fatal heart attacks occurred as well as 115 deaths due to coronary disease. Those nurses in the top 20 per cent of dietary vitamin E intake (as calculated from the questionnaire) showed 23–50 per cent less heart disease after correction of the data for age and smoking. Women who took vitamin E supplements for short periods had little apparent benefit but those who took them for more than two years had a relative risk of major coronary heart disease of 0.38–0.91[†] after adjustment for age, smoking, risk factors, and intake of other antioxidant nutrients.

The authors concluded that **'although these prospective data do not prove a cause and effect relationship, they suggest that among middle-aged women the use of vitamin E supplements is associated with a reduced risk of coronary heart disease.'** They also recommended that **'public policy recommendations about the widespread use of vitamin E should await the results of more definitive trials.'**

Second paper

In 1986, 39 910 US male health professionals (dentists, veterinary surgeons, pharmacists, opticians, osteopaths, podiatrists) 40–75 years old, free of obvious coronary disease, completed detailed dietary questionnaires designed to assess their intake of vitamin C, carotene, and vitamin E. During four years of follow-up, 667 cases of coronary disease were observed. After controlling for age and other risk factors, a lower risk for coronary disease was observed among men with apparently high vitamin E intakes. For men consuming more than 60 International Units per day the relative risk was 0.49–0.83 compared to those eating less than 7.5 units per day. As

[†] Control risk is 1; a value below 1 means a decreased risk. Thus, they had 38–91% of the control risk.

compared with men who did not take vitamin E supplements, men who took at least 100 units per day for at least two years have a relative risk of 0.47–0.84 for coronary disease. By contrast, β-carotene intake was not associated with a lower risk among those who had never smoked, but among smokers, high β-carotene intake did appear to associate with less risk. A high intake of vitamin C was not associated with a lower risk of coronary disease. The authors concluded that **'These data do not prove a causal relation but they provide evidence of an association between a high intake of vitamin E and a lower risk of coronary heart disease in men.'** They also commented, as did the authors of the first paper, that **'public policy recommendations with regard to the use of vitamin E supplements should await the results of additional studies.'**

Comments

The authors of both papers were very careful to point out the limitations of their results, but their cautions were often ignored in the lay press (Figure A1). What are the problems? Here are some obvious ones:

(1) Questionnaires about food/vitamin intake are not substitutes for direct measurement of body levels. The correlation between vitamin E intake as assessed by this type of questionnaire and blood vitamin E levels is less than 50 per cent. This also makes it very difficult to control adequately for the effects of other antioxidants, such as carotene and vitamin C (again, intake only correlates poorly with body level). Comparison of nutrient intake as specified on the questionnaire with weekly diet records in 127 subjects gave poor correlations for vitamin E in subjects not consuming supplements (i.e. the errors in estimating vitamin E level by questionnaire in non-supplementing subjects are likely to be large). This means that real associations between vitamin E intake from the diet are likely to be missed, and the effects of supplements thereby are emphasized (it is easy to remember if you take pills or not).

(2) People who consume vitamin E are more likely to be 'health

aware' and to have other aspects of their lifestyle (attention to diet, taking exercise, etc.) that could decrease heart disease. (Perhaps healthier subjects select themselves for supplementation.) However, if self-selection were the entire explanation, it is difficult to explain why vitamin C did not appear to be protective.

(3) Plasma cholesterol levels were not measured. High cholesterol is an important risk factor for heart disease (see Chapter 7).

(4) Other diseases were not examined. We cannot be totally sure that perhaps vitamin E protects against heart disease but does not increase death from other causes.

(5) In a European multicentre study (the EURAMIC study) published in *The Lancet* (volume 342, page 1379, 4 December 1993), vitamin E and β-carotene were measured in human fat tissue. There was no difference in vitamin E between people with heart attacks or controls. Fat vitamin E levels were chosen as a measure of long-term status, presumably much more accurate than dietary questionnaires. Low β-carotene levels were associated with heart disease only in patients who smoked.

(6) By contrast in an earlier study from Scotland (*The Lancet*, volume 337, page 1, 5 January 1991) levels of vitamin E in blood plasma were found to be lower in patients with angina pectoris (chest pain that occurs when blood flow to the heart is decreased due to artery disease). Levels of β-carotene and vitamin C were also lower in patients, but this effect was largely due to the fact that they smoked more than controls.

The Linxian study

The rural county of Linxian, China, has a high death rate from cancer of the oesophagus and stomach. A clinical trial of vitamin supplementation has been published (*Journal of the National Cancer Institute*, volume 86, page 1483, 15 September 1993). Subjects aged 40–68, 29 584 in total, were given various nutritional supplements or an inert pill (a placebo) of the various supplements. It was found that a mixture of selenium, β-carotene, and vitamin E produced significant falls in death from stomach cancer after 5 years, whereas supplementation with a vitamin C-containing mixture did not.

Comments
The Linxian population was poorly nourished to start with (but then so are, on average, the poorest subgroups of the population of the USA and UK). The effect on cancer mortality is convincing, but it is, of course, uncertain which component(s) of the supplement were responsible.

Authors' overall conclusions
It is easy to 'pick holes' in epidemiological studies. Questionnaires about diet are a very approximate way of measuring intake of antioxidants. But is the vitamin E content of a piece of body fat any better? Or blood plasma vitamin E levels? It is difficult to be certain. Overall, it seems that a high intake of vitamin E (perhaps involving the use of supplements) may offer some protection against cardiovascular disease. β-Carotene may protect against cancer, but only seems to affect cardiovascular disease in smokers. Vitamin C seems not to affect cardiovascular disease much. Thus, avoiding smoking and maintaining a good intake of fruits and vegetables, rich in antioxidant nutrients, would seem to be the key to human health, perhaps combined with moderate vitamin E supplementation (see the conclusion on page 134).

However, just to complicate things further, here is the Finnish study.

The Finnish study
The Finnish study, published in the *New England Journal of Medicine* (volume 330, pages 1029–1035) in April 1994, has attracted a great deal of attention as apparently showing a deleterious effect of β-carotene supplementation. The study examined 29 133 male smokers (age range 50–69 years, mean age 57 years) who had smoked (on average) 20 cigarettes a day for 36–37 years. Some

received 20 mg of β-carotene per day, others 50 mg of vitamin E (dl-α-tocopherol), some both, and some placebos. They were followed for 5–8 years and the incidence of lung cancer noted. Vitamin E had no significant effect. However, after 3 years of the study, subjects receiving β-carotene showed a significantly greater incidence of lung cancer than controls.

Comments

Previous studies have shown that low blood β-carotene levels are associated with higher levels of lung cancer (and some other forms of cancer). The obvious question is: is the β-carotene responsible or is high blood β-carotene simply a 'marker' of a diet rich in fruits and vegetables? This direct test of the effect of β-carotene would seem to suggest that it is not protective and may be deleterious. However, patients who have smoked for 36–37 years are already well on the way to developing various pathologies, including lung cancer and it is unlikely that β-carotene or vitamin E could reverse this. Hence, a logical result of this trial would be 'no effect' as was seen with vitamin E for lung cancer. It remains to be explained why β-carotene increased deaths from lung cancer.

It could be due to the following.

1. A random statistical 'fluke'.
2. β-Carotene at high doses is toxic or interacts with cigarette smoke to produce toxic products.
3. The β-carotene was contaminated with possibly toxic breakdown products.
4. The smokers receiving β-carotene had, on average, smoked a year longer than controls. In patients at high risk of developing lung cancer, this could be enough to skew the results. In any case, the results suggest that the best way of protecting against smoke-induced pathology is to give up smoking rather than to keep smoking and take antioxidants.

5 Oxidative stress

The body is a conglomeration of chemical matters; when these are deranged, illness results, and nought but chemical medicines may cure the same.

Theophrastus Paracelsus (1527)

Oil easily combines with oxygen. This combination is either slow or rapid. In the first case rancidity is the consequence, in the second inflammation.

Chaptal (1791)

Introduction

Oxygen radicals and hydrogen peroxide are constantly produced in the human body. Some of this production is a chemical accident, such as generation of hydroxyl radicals (OH$^\bullet$) by our constant exposure to low levels of radiation from the environment (see Chapter 1) and of superoxide (O$_2^{\bullet-}$) by leakage of electrons from electron transport chains (see Chapter 2). Other production of these species is deliberate, including the generation of H$_2$O$_2$ by such enzymes as D-amino acid oxidase (see Chapter 2). Perhaps the best-known examples of deliberate production of radicals in the human body are the generation of O$_2^{\bullet-}$ by activated phagocytic cells (see the Appendix to this chapter) and the generation of nitric oxide by the cells lining blood vessel walls (see Chapter 1).

We have seen that the potentially damaging effects of O$_2^{\bullet-}$ and H$_2$O$_2$ are diminished by the action of antioxidant defence systems (see Chapter 3). It is difficult to defend against background ionizing radiation causing the splitting of water and generating OH$^\bullet$: the

OH˙ is so reactive that it will attack whatever is present at its site of formation in the body. Instead, the damage has to be repaired (Table 1). In addition, it seems that the human body has just sufficient antioxidant defences to balance the normal rate of production of oxygen-derived species: there is no great reserve of antioxidant defence capacity. Perhaps one reason for this is that some oxygen-derived species are useful.

Because generation of oxygen-derived species and the level of antioxidant defence systems are approximately balanced, it is easy to tip the balance in favour of the oxygen-derived species and upset cell biochemistry. This imbalance is called *oxidative stress*. Most cells can tolerate a mild degree of oxidative stress because they have *repair systems* which recognize and remove oxidatively damaged molecules (Table 1), which are then replaced. In addition, cells may increase antioxidant defences in response to the stress. For example, placing rats in an atmosphere of pure (100 per cent) O_2 (normal air is 21 per cent O_2) kills them within a few days. However, exposing the animals to gradually increasing O_2 concentrations over a period of days enables them to increase the activity of antioxidant defences in their lungs, so that they can eventually tolerate 100 per cent O_2. However, severe oxidative stress can injure or kill cells.

How can oxidative stress be imposed?

There is normally a balance between production of oxygen-derived species and antioxidant defences. It follows that there are at least two ways of imposing oxidative stress: decrease the antioxidants or increase the production of oxygen-derived species so that the antioxidants can no longer cope. A third way is to increase the availability of transition metal ions, so that a greater fraction of the O_2˙ and H_2O_2 produced normally is converted into damaging OH˙ radicals.

Depletion of antioxidants

Antioxidant defences are highly dependent upon adequate nutrition (see Chapter 4, Table 1) and so malnutrition can lead to oxidative

Table 1. *Repair systems that can deal with free radical damage in the human body*

Molecule damaged by oxygen-derived species	Type of damage	Repair mechanism
DNA	Changes in the purine (adenine, guanine) or pyrimidine (cytosine, thymine) bases. Attack on the sugar (deoxyribose). Breakage of the DNA backbone in one strand (single-strand breaks) or both strands (double-strand breaks) of the DNA double helix.	The DNA is 'patrolled' by repair enzymes. When an error is detected, the portion of DNA containing it is removed and the correct DNA inserted by enzymes that use the information in the other strand of the double helix to tell them what to insert. Double-strand breaks prevent this and so are more damaging to cells than single-strand breaks.
Proteins	Breakage of protein backbone, oxidation of thiol ($-SH$) groups and of other amino acid residues, cross-linking of different protein molecules by joining together of amino acid radicals on different protein molecules.	Special enzymes recognize 'abnormal' proteins and destroy them.
Lipids	Damage to membrane-bound proteins and oxidation of PUFA side chains during lipid peroxidation.	Normal 'turnover' of membrane components. A special glutathione peroxidase may remove peroxidized fatty acids from membranes (Chapter 3).

stress. Some human diseases appear to result from inadequate intake of antioxidant nutrients, such as the neurodegeneration resulting from prolonged deficiency of vitamin E in patients unable to handle fats properly (see Chapter 4). In patients infected with human immunodeficiency viruses, the viruses that cause AIDS (acquired immunodeficiency syndrome), there are reports of abnormally low concentrations of reduced glutathione (GSH) in their lymphocytes, cells important in the immune response. It has been speculated that these lower levels of GSH may render the cells more prone to oxidative stress, thus perhaps contributing to the progressive loss of certain classes of lymphocyte that is a hallmark of AIDS. **Table 2 reviews current knowledge of this complex area; our knowledge is incomplete and certainly does not justify the aggressive marketing of 'antioxidant nutritional supplements' to AIDS patients seen in some parts of the USA.**

Increased formation of oxygen-derived species

More usually, however, oxidative stress results from increased *production* of oxygen-derived species such as $O_2^{\cdot-}$, H_2O_2, and $OH^{\cdot}$. *Increased O_2 concentration* is one mechanism for achieving this, as we have seen in Chapter 1 (in relation to the superoxide theory of oxygen toxicity). *Inappropriate activation of phagocytes*, as in several diseases involving chronic inflammation, exposes the surrounding tissues to $O_2^{\cdot-}$, H_2O_2 and $NO^{\cdot}$, as well as to other potentially noxious materials that activated phagocytes secrete, such as hypochlorous acid (see the Appendix to this chapter). The metabolism of various *drugs and toxins* may impose oxidative stress. First, many drugs and toxins are prepared for excretion from the body by being joined together with GSH, in reactions catalysed by *glutathione transferase* enzymes. Large doses of toxin can deplete cells of GSH, rendering them more sensitive to damage by oxygen-derived species.

Second, some drugs are converted into free radicals. Often the *cytochrome P450 system* of the cell's endoplasmic reticulum is responsible. Cytochrome P450 is a collective name given to a group of cytochromes present within the endoplasmic reticulum of many

Table 2. *Oxidative stress and AIDS—There is much speculation in the medical journals that oxidative stress is involved in the development of AIDS in patients infected with HIV viruses and that antioxidants might be beneficial. When the viruses first infect cells, they insert DNA (coding for more virus) into the cell DNA. The viral DNA lies dormant for a long time, but eventually activates.*

Early after HIV infection, there is a fall in the GSH content of some body fluids and some lymphocytes.

This fall in GSH *could* interfere with lymphocyte function—there is as yet no direct evidence that it does.

In patients who have developed AIDS, malnutrition *may* occur because of loss of appetite and/or impaired absorption of food from the gut. Malnutrition can limit the intake of dietary antioxidants.

AIDS patients produce increased levels of a protein; *tumour necrosis factor*, that affects many body cells and may increase intracellular production of oxygen-derived species within cells. Most cells respond to this by increasing antioxidant defences, but HIV-infected cells *might* be unable to do so.

Experiments with isolated HIV-infected cells have shown that expression of the virus can be stimulated by H_2O_2 and inhibited by thiols, including compounds that raise the GSH content of cells.

Some of the drugs used in treatment of AIDS (especially AZT) do considerable damage to mitochondria, which *may* cause increased leakage of electrons and more $O_2^{\cdot-}$ formation.

(Note how uncertain our knowledge is at present.)

cells. Their function is to detoxify 'foreign compounds'. The highest concentration of cytochromes P450 in the human body is in the liver, which has to detoxify many foreign compounds. Some lung cells also contain high levels of P450. The P450 enzyme systems have evolved to detoxify many foreign compounds by converting them into products that are more soluble in water and so are excreted more easily from the body. In a few cases, however, the products are more dangerous than the starting material.

For example, the cancer-causing (carcinogenic) hydrocarbon

benzpyrene, produced by burning organic material such as high temperature barbecuing of meat and a contributor to the cancers caused by cigarette smoking, is metabolized by P450s (assisted by other enzymes) to yield more-reactive products that attack DNA and are the real carcinogens. Another example is carbon tetrachloride. This organic solvent is converted by cytochromes P450 into a free radical. A carbon–chlorine covalent bond is broken to leave an unpaired electron on the carbon.

$$CCl_4 \overset{P450}{\to} {}^{\cdot}CCl_3 + Cl^- \tag{1}$$

The ${}^{\cdot}CCl_3$ reacts fast with O_2 to give a peroxyl radical.

$$^{\cdot}CCl_3 + O_2 \to CCl_3O_2{}^{\cdot} \tag{2}$$

Like the peroxyl radicals formed during lipid peroxidation (see Chapter 3), $CCl_3O_2{}^{\cdot}$ is very good at abstracting hydrogen from polyunsaturated fatty acid side chains in membrane lipids, so starting lipid peroxidation. Thus a major mechanism of CCl_4 toxicity is to induce uncontrolled peroxidation of the lipids of the cell's endoplasmic reticulum, a site of important metabolic reactions, including the synthesis of proteins. The liver is the major organ affected in CCl_4 poisoning. Damage is worse if the tissues are low in vitamin E. Thus an inadequate intake of antioxidant nutrients can increase the damaging effects of drugs, toxins, and environmental pollutants. The levels of P450 are also important: these rise after consuming many drugs and after frequent intake of large amounts of alcohol. This is a protective response but unfortunately a few toxins (such as CCl_4) are made more dangerous thereby.

Acetaminophen (sometimes called paracetamol) is another example of a toxin that promotes oxidative stress. In approved doses, it is a safe painkiller. Acetaminophen is metabolized by cytochromes P450 into a product that combines with GSH. Overdoses of acetaminophen severely deplete GSH, especially in the liver, and can lead to liver failure. If treatment is given early enough, the consequences of acetaminophen overdosage can be prevented by keeping liver GSH levels high. GSH itself is not absorbed easily by cells, but the drug *N-acetylcysteine* is quickly absorbed and provides the amino acid cysteine inside the cell, which is used to make GSH.

Redox-cycling drugs

Some other drugs and toxins undergo *redox cycling* to generate $O_2^{\bullet-}$. In redox cycling, the compound is reduced by the body to a free radical. This radical reacts with O_2 to regenerate the compound and make $O_2^{\bullet-}$. Hence the redox-cycling compound effectively *catalyses* reduction of O_2 to $O_2^{\bullet-}$.

Many drugs and toxins that redox cycle are quinones (Figure 1). Quinones can be reduced by enzymes inside the body to semiquinone radicals (Figure 1), which react with O_2 to regenerate the quinones and produce $O_2^{\bullet-}$. It has also been claimed that several semiquinones and hydroquinones are able to reduce iron attached to the storage protein ferritin and release it from the protein, so making iron, catalytic for free radical reactions, more available. Quinones are widely distributed in nature, often being present in plants and insects as deterrent chemicals to protect against predators. Perhaps the best example of how a quinone can be used effectively by an insect is seen in the bombardier beetle (Figure 2). When a bombardier beetle is disturbed, it sprays a pulsed jet (500 times per second) of hot solution containing hydroquinones and hydrogen peroxide. A sac near the insect's 'ejection mechanism' contains the concentrated solution of H_2O_2 plus hydroquinone. To activate the

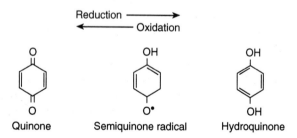

Figure 1. Quinones, hydroquinones, and semiquinone radicals. Many drugs and toxins contain quinone or hydroquinone rings, including the anti-cancer drug adriamycin.

Figure 2. A bombardier beetle in action. (Photograph courtesy of Thomas Eisner and Daniel Aneshansley.)

spray, the contents of the sac are pushed into a reaction chamber containing catalase and peroxidase. The hydroquinones are oxidized to quinones by the peroxidase, and the heat and pressure produced by H_2O_2 breakdown to O_2 gas by catalase, cause the temperature to rise close to 100 °C—a perfect 'anti-mugging' spray and yet another illustration of a useful role for reactive oxygen species!

One example of a quinone is the anti-cancer drug doxorubicin (sometimes called adriamycin), which is frequently (and effectively) used in the treatment of leukaemias. It is thought to act by stopping the cancer cells from making new DNA. This prevents them from dividing. Like most powerful drugs, adriamycin produces various side-effects, the most severe being heart damage. This *cardiotoxicity* is thought to be mediated by a reduction of the adriamycin molecule (ADR) by enzymes in the heart, producing an adriamycin radical (ADR$^{\cdot-}$)

$$ADR + e^- \rightarrow ADR^{\cdot-} \tag{3}$$

The adriamycin radical reacts very fast with O_2,

$$ADR^{\cdot-} + O_2 \rightarrow ADR + O_2^{\cdot-} \tag{4}$$

and so the cell can be 'swamped' with $O_2^{\cdot-}$. Adriamycin is also an excellent binder of iron, and so may provide iron to catalyse conversion of $O_2^{\cdot-}$ and H_2O_2 (made from the $O_2^{\cdot-}$) into $OH^{\cdot}$ radicals. Adriamycin, like all redox-cycling drugs, catalyses the reduction of O_2 to $O_2^{\cdot-}$. Other examples of redox-cycling toxins are menadione and paraquat (Table 3).

Smoking imposes an oxidative stress on the lung and probably on most other body tissues, since both the tar component and the gas component of cigarettes are rich in radicals (Table 4).

Table 3. *Some redox cycling drugs*

Name	Nature	Biological effects
Menadione	A quinone	Toxic to many cells, including liver and red blood cells. Often used in the laboratory to impose oxidative stress upon isolated cells for experimental purposes
Paraquat	A herbicide, used as a weed killer	Selectively accumulated by the lung, causes severe lung damage
Alloxan	Used to make animals diabetic, for research purposes	Undergoes redox cycling in the insulin-producing cells of the pancreas, so destroying them and causing insulin-deficiency diabetes

Table 4. *Why cigarette smoking may impose oxidative stress*

Smoke contains many free radicals (both in the gas and tar phases), especially peroxyl radicals, that might attack biological molecules and deplete antioxidants such as vitamins C and E.

Smoke contains oxides of nitrogen, including the unpleasant nitrogen dioxide ($NO_2{}^\bullet$).

The tar phase of smoke contains hydroquinones. These may enter the tissues and redox cycle. Some may also release iron from ferritin.

* Smoking may irritate lung macrophages, encouraging them to make $O_2{}^{\bullet-}$.
* Smokers' lungs contain more neutrophils than the lungs of non-smokers, and smoke might activate these cells to make $O_2{}^{\bullet-}$.

Smokers often eat poorly and drink more alcohol than non-smokers and may have an insufficient intake of nutrient antioxidants.

---

* The effects of cigarette smoke on phagocytes are dose-related. Low levels may stimulate them, but high levels may poison them and so depress their activity. For an explanation of phagocyte types, see the Appendix to this Chapter.

The consequences of oxidative stress
Calcium metabolism

Cells can tolerate mild oxidative stress, and often respond by raising their levels of antioxidant defences. However, severe oxidative stress produces serious disturbances in cell metabolism. The intracellular levels of calcium (Ca^{2+}) ions, important stimulators of many cell processes, are very low but transiently increase in response to several hormones, producing an important metabolic signal. Oxidative stress can damage the proteins that keep the Ca^{2+} low, so causing *prolonged* and *excessive* rises in intracellular Ca^{2+}. If the Ca^{2+} level rises too high, it can activate enzymes that attack DNA (nucleases), cutting the backbone and so fragmenting the DNA. High Ca^{2+} can also activate enzymes that cleave structural proteins within the cell, causing the membrane to 'bleb out' (Figure 3). Oxygen-derived species, if not adequately scavenged, may also

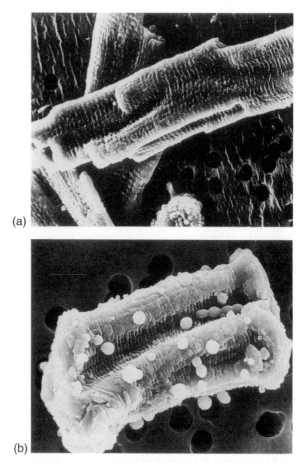

(a)

(b)

Figure 3. Membrane blebbing in heart cells. The electron photomicrographs show (a) normal heart cells and (b) cells after treatment with a synthetic lipid hydroperoxide. Formation of membrane blebs on the cell surface is clearly shown. Courtesy of Dr A.A. Noronha-Dutra.)

oxidize thiol (−SH) groups on cellular structural proteins and contribute to blebbing. If blebbing goes too far, the bulge may rupture, producing a hole in the membrane that kills the cell.

Why do cells have these potentially lethal DNA-cutting enzymes? As it grows and changes shape, the body needs to be able to destroy certain cells, including cells in the immune system that recognize normal body tissues as foreign organisms. If these cells were allowed to remain, they would attack normal body tissues, so causing *autoimmunity*. The programmed death of cells appears to involve increases in Ca^{2+} and resulting DNA fragmentation in response to certain hormones. Some scientists believe that severe oxidative stress can activate this self-destruction mechanism in normal cells.

DNA damage

DNA fragmentation is frequently seen in cells subjected to oxidative stress. Ca^{2+}-dependent nucleases may be involved. Another possibility is that hydroxyl radicals ($OH^{\cdot}$) are formed in the nucleus and attack the DNA. DNA and/or certain proteins adjacent to it may contain bound metal ions that catalyse $OH^{\cdot}$ formation—copper and iron are likely candidates. In addition, oxidative stress may cause release of iron ions within the cell—these ions could bind to DNA, so making it a target of attack by $OH^{\cdot}$ generated at the site of iron binding. For example, it has been suggested that when mitochondria take up free Ca^{2+}, they may release Fe^{2+} within the cell. If DNA has bound Fe^{2+} or Cu^{+}, then any H_2O_2 reaching the nucleus can damage DNA by making $OH^{\cdot}$.

Excessive DNA fragmentation, whether by nucleases or by attack by $OH^{\cdot}$, leads to activation of the enzyme *poly(ADP-ribose) synthetase* within cells. This enzyme acts upon NAD^{+}, which plays a key part in cell function (see Chapter 2, Figure 2). It splits the NAD^{+} molecule and attaches part of it to proteins in the nucleus, possibly to help the DNA repair process. If there are many DNA strand breaks, the synthetase enzyme may use up so much NAD^{+} that cell metabolism halts, and the cell dies. This *lethal NAD^{+} depletion* has been suggested to be a mechanism by which cells with serious DNA damage are eliminated by the body, presumably because large amounts of DNA repair might lead to mutation, since DNA repair enzymes have a small (but finite) chance of making mistakes and causing mutations.

Lipid peroxidation

Some toxins act mainly by stimulating lipid peroxidation in cells and destroying cell membranes—one example is carbon tetrachloride. The increased availability of intracellular iron that may result from oxidative stress can also promote lipid peroxidation. Lipid peroxidation not only damages lipids but also membrane proteins such as receptors and enzymes.

Protein damage

Membrane-associated proteins can be damaged during lipid peroxidation. Proteins can also be damaged directly by free radicals. In particular, OH$^{\cdot}$ attacks most amino acid residues on proteins. Thiol (−SH) groups are particularly sensitive to attack by a range of free radicals. Many proteins bind copper and iron ions; this can make the protein a target of attack by OH$^{\cdot}$ if H_2O_2 is then generated. The OH$^{\cdot}$ is generated upon the protein at the site of the bound metal ion and then damages that site. However, the human plasma iron and copper transport proteins (transferrin and ceruloplasmin) do not allow such reactions to occur with their bound metal ions. By contrast, copper ions bound to plasma albumin will react with H_2O_2 to form OH$^{\cdot}$, so damaging the albumin. However, the albumin can quickly be replaced by the liver, so that this damage may be of little consequence (discussed in Chapter 2).

Which is the most important cell injury process mediated by oxidative stress?

The simple answer to the above question is 'it depends'. The relative importance of different targets of injury varies enormously depending on the tissue studied, the means used to impose oxidative stress and the degree of such stress employed. Thus there is no single mechanism by which oxidative stress kills or injures cells, and no single antioxidant will protect against every type of damage.

Appendix to Chapter 5
Human phagocytes: useful dealers of death and destruction

Stimulate the phagocytes. Drugs are a delusion.
The Doctors' Dilemma G. Bernard Shaw (1856–1950)

All substances are poisons: there is none which is not a poison. The right dose differentiates a poison and a remedy.
Theophrastus Paracelsus (1493–1541)

The production of $O_2^{\cdot -}$ and H_2O_2 by some types of white blood cell is one example of a useful role for these oxygen-derived species, but it can become a source of tissue damage if their production is too widespread or goes on for too long.

Human blood contains large numbers of *neutrophils* (about 2.5–7.5 million per ml in healthy subjects), amoeba-like white blood cells whose function is to recognize, engulf, and destroy foreign organisms, such as bacteria and viruses (Figure A1). This process of engulfment is called *phagocytosis*, and cells which can do it are often called *phagocytes*. Indeed, infection usually causes a rise in the 'white blood cell count' (number of cells per ml of blood) as the body mobilizes more of these cells to fight the infection.

Recognition of a foreign organism by a neutrophil usually occurs because the foreign organism has been coated by an antibody protein produced by the immune system of the body. As soon as an organism recognizable as foreign touches the surface of the neutrophil, an enzyme in the membrane of this cell is activated (Figure A2). This enzyme removes electrons from NADPH inside

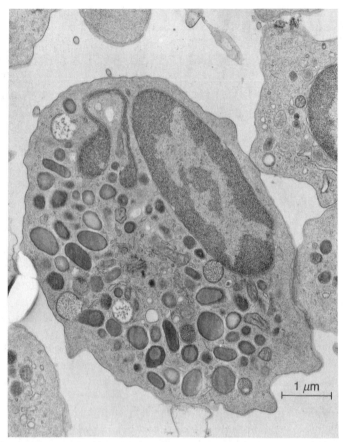

Figure A1. Amoeba-like structure of a human neutrophil as seen under the electron microscope. (Courtesy of Dr David Hockley.)

the cell, oxidizing it to $NADP^+$. The electrons are passed across the membrane and used to reduce O_2 to $O_2^{\cdot-}$ on the outer surface of the neutrophil. Hence when the foreign organism is engulfed by the neutrophil, it is wrapped up in a piece of membrane generating $O_2^{\cdot-}$. This $O_2^{\cdot-}$ production is one of the mechanisms by which

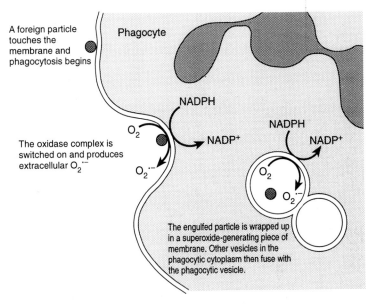

Figure A2. A diagrammatic representation of phagocytosis and oxygen radical production by neutrophils. Contact with foreign particles activates an oxidase enzyme in the cell membrane of the neutrophil, which produces superoxide radical ($O_2^{\cdot-}$) on the outside of the membrane. (Courtesy of the Upjohn Co., 'Current Concepts' series.)

engulfed foreign organisms are killed by phagocytes. Another killing mechanism sometimes used is the production of toxic amounts of nitric oxide, $NO^{\cdot}$. The microbial killing process ultimately leads to the death of the phagocytic cells themselves from oxidative stress and other processes.

The importance of $O_2^{\cdot-}$ production by neutrophils to human health is simply illustrated by examining patients whose neutrophils lack the ability to do it—the inborn disease called *chronic granulomatous disease* (Table A1). Patients with this disease suffer persistent infections with certain bacteria, presumably those whose killing by neutrophils is largely superoxide dependent.

Table A1. *Chronic granulomatous disease (CGD)*

Prevalence in humans	1 in 1 000 000
Age of onset	First symptoms at or before 1 year of age in 80% of cases
Clinical manifestations	(1) Recurrent bacterial and fungal infections (2) Consequences of chronic inflammation (failure to thrive, anaemia, enlarged spleen, etc.) (3) Granuloma formation
Diagnosis	Absence of $O_2^{\cdot-}$ and H_2O_2 production by activated phagocytes
Organ involvement	(1) Lung (80%) (2) Lymph nodes (75%) (3) Skin (65%) (4) Liver (40%) (5) Bone (30%) (6) Gastrointestinal (30%) (7) Sepsis and meningitis (20%) (8) Genitourinary (10%)
Prognosis	Variable: death in infancy to middle age
Treatment	Antibiotics

Blood also contains smaller numbers of other kinds of phagocytes, called *eosinophils* and *monocytes* (Table A2). Like neutrophils, eosinophils and monocytes produce $O_2^{\cdot-}$ upon activation. Eosinophils appear particularly important in dealing with parasites other than infectious bacteria, and their number in the blood is increased in patients with allergies. Monocytes are important at sites of inflammation (Figure A3): they enter inflamed sites after the neutrophils and develop into *macrophages* (Figure A4). Both monocytes and macrophages produce $O_2^{\cdot-}$, whereas production of $NO^{\cdot}$ as a toxic agent may be more important in the case of macrophages.

Table A2. *White blood cells in human blood*

Type	Number per ml of blood	Function
Neutrophils	2.5–7.5 million	Recognize, engulf, and destroy foreign organisms
Eosinophils	0.04–0.4 million	Anti-parasite, involved in allergies
Monocytes	0.2–0.8 million	Precursors of macrophages
Lymphocytes	1.5–3.5 million	Several different types, involved in the immune response. Some types of lymphocytes are attacked by the HIV viruses that cause AIDS

The Killing Mechanisms

Although the symptoms of chronic granulomatous disease tell us that $O_2^{\cdot-}$ is important in the killing of bacteria by phagocytes, the mechanism by which killing occurs is uncertain. Superoxide is not very toxic and it cannot get inside bacteria. It can form H_2O_2

$$2O_2^{\cdot-} + 2H^+ \rightarrow H_2O_2 + O_2 \tag{A1}$$

which is toxic to many bacterial strains. The H_2O_2 easily enters the bacterium (since it resembles water) and reacts with iron or copper inside the bacteria to produce damaging $OH^\cdot$. Superoxide could also react with $NO^\cdot$ (if both were generated at the same time) to give peroxynitrite

$$O_2^{\cdot-} + NO^\cdot \rightarrow ONOO^- \tag{A2}$$

Peroxynitrite itself might be damaging to bacteria. It might also break down to give $OH^\cdot$

$$ONOO^- + H^+ \rightarrow OH^\cdot + NO_2^\cdot \tag{A3}$$

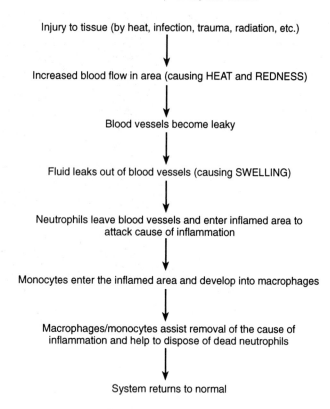

Injury to tissue (by heat, infection, trauma, radiation, etc.)

Increased blood flow in area (causing HEAT and REDNESS)

Blood vessels become leaky

Fluid leaks out of blood vessels (causing SWELLING)

Neutrophils leave blood vessels and enter inflamed area to attack cause of inflammation

Monocytes enter the inflamed area and develop into macrophages

Macrophages/monocytes assist removal of the cause of inflammation and help to dispose of dead neutrophils

System returns to normal

Acute inflammation is a normal response to injury. In some diseases, such as *rheumatoid arthritis* and the *inflammatory bowel diseases*, persistent inflammation leads to severe tissue damage.

Figure A3. Sequence of events in 'normal' acute inflammation in humans.

Neutrophils, monocytes, and eosinophils (but not macrophages) also contain peroxidase enzymes. The best-characterized is that of neutrophils, a green-coloured protein called *myeloperoxidase*. The colour is due to a haem-like grouping at the active site of the

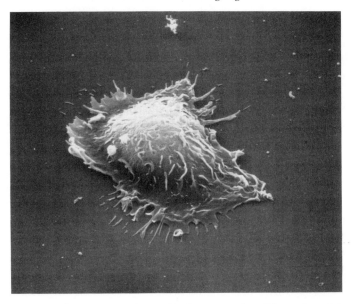

Figure A4. A human macrophage adhering to a surface. The surface of the cell is shown, as seen under the electron microscope. (Courtesy of Dr David Hockley.)

enzyme. Activated neutrophils can release some of this enzyme, causing fluids generated at sites of infection to have a greenish tinge—hence the green colour of pus and of sputum coughed up during chest infections. Myeloperoxidase is classified as a 'non-specific' peroxidase (see Chapter 3)—it can use H_2O_2 to oxidize a wide range of products. However, it is believed that, in the body, its major substrates are chloride (Cl^-) and bromide (Br^-) ions, especially the former. Myeloperoxidase uses H_2O_2 to catalyse the reaction

$$H^+ + Cl^- + H_2O_2 \rightarrow HOCl + H_2O \tag{A4}$$

HOCl, *hypochlorous acid*, is a weak acid and at pH 7.4 half of it exists as the hypochlorite ion OCl^-.

$$HOCl \rightarrow H^+ + OCl^- \tag{A5}$$

HOCl and OCl⁻ are very damaging to bacteria—indeed many disinfectants and bleaches used in the home are hypochlorite solutions, usually sodium hypochlorite, NaOCl. HOCl is not a free radical, since it contains no unpaired electrons.

Thus activated phagocytes can kill bacteria by imposing severe oxidative stress upon them. The phagocytes can also damage themselves at the same time—activated neutrophils at sites of inflammation rapidly become 'spent' and disintegrate or are removed by being phagocytosed by macrophages (Figure A3). One problem with phagocytes is that their activation imposes oxidative stress on the surrounding tissues as well and, if too many phagocytes are activated or if inflammation goes on for too long, serious injury can result. Examples of such injury are damage to the joints as a result of chronic joint inflammation in the disease *rheumatoid arthritis*, and damage to the large intestine (often leading to cancer) as a result of chronic inflammation of this organ in the *inflammatory bowel diseases*.

6 Free radicals and antioxidants in ageing and disease: fact or fantasy?

Age will not be defied.

Of regiment of health Francis Bacon (1561–1626)

Every day you get older; it's a law.

Butch Cassidy in *Butch Cassidy and the Sundance Kid* (1972)

It is difficult these days to open a popular science magazine or medical journal without seeing an article about the role of free radicals in human disease. Women's magazines contain articles about the use of 'antioxidant creams' to prevent ageing of the skin, and health food stores are full of pamphlets and books describing how antioxidant or dietary supplements can prolong life, protect youth, or prevent disease. Is there any truth in all this?

Three points are worth stating:

(1) There is no evidence that consuming more antioxidants will prolong the natural maximum human life span.
(2) There is reasonable evidence that eating diets rich in fruits and vegetables decreases the incidence of cancer and heart disease (see Chapters 4 and 7). Antioxidants present in such diets (including ascorbic acid, vitamin E, and carotenoids) *may* cause this effect, or contribute to it (see the Appendix to Chapter 4). Other factors may be equally or more important, however (*See*

Chapter 4, Table 4). For example, whole-grain cereals contain natural chelating agents (such as *phytic acid*) that might diminish iron uptake by the human gut. Some authors have suggested that high body iron stores may predispose to cancer and heart disease.

(3) **Free radicals are formed in greater amounts during all human diseases (i.e. some degree of oxidative stress probably occurs in all diseases). This does *not* mean that free radicals cause disease, or that they contribute significantly to its course.** Similarly, a number of toxins cause oxidative stress (Table 1), but it does not mean that this is the major mechanism by which they cause damage.

Let us explore these principles in more detail.

Oxidative stress as a cause of human tissue injury

What is the exact role played by oxygen-derived species in human disease? Some human diseases may be *caused* by oxidative stress. Thus, γ-rays generate $OH^\bullet$ by splitting water molecules (see Chapter 1):

$$H_2O \rightarrow OH^\bullet + H^\bullet \tag{1}$$

Many of the biological consequences of excess radiation exposure may be due to free radical damage to proteins, DNA, and lipids, since $OH^\bullet$ attacks all these molecules. The signs produced by chronic dietary deficiencies of selenium (e.g. Keshan disease) or of tocopherols (neurological disorders seen in patients with defects in intestinal fat absorption) might also be mediated by oxidative stress (see Chapter 4). In the premature infant, exposure of the undeveloped retina to elevated concentrations of oxygen can lead to *retinopathy of prematurity*, which in most severe forms can result in blindness. Several clinical trials have documented the efficiency of α-tocopherol in minimizing the severity of the retinopathy, *consistent with* (but by no means proving) a role for lipid peroxidation. Another suggestion is that the risk of bleeding within the brains of premature infants may be minimized by giving α-tocopherol. Free

Table 1. *Some toxins that impose oxidative stress in human tissues*

Toxin	Consequences
Cigarette smoke	Rich source of free radicals (Chapter 5, Table 4). May contribute to destruction of lung elastic fibres (causing *emphysema*) and to lung cancer.
Asbestos	Asbestos fibres contain iron and catalyse conversion of H_2O_2 to $OH^\bullet$. It is speculated that this contributes to lung injury and lung cancer induced by this toxin.
Alcohol	Excess alcohol intake may deplete vitamin E and GSH in cells and increase iron uptake from the gut.
Adriamycin	A quinone drug, used to treat leukaemias. Its major side-effect, heart damage, may involve free radicals (Chapter 4).
Carbon tetrachloride	CCl_4 causes severe liver damage: lipid peroxidation is involved (Chapter 4).
GSH-depleting agents	Anything that depletes GSH can make cells more prone to oxidative damage. Examples of drugs that can do this are acetaminophen (Chapter 4) and cocaine.
Metal ions	Poisoning by excess intake of iron, copper, manganese, nickel, cadmium, chromium, and lead *may* involve GSH depletion and free radical generation.
Air pollutants	Ozone (O_3) and nitrogen dioxide ($NO_2^\bullet$) are powerful oxidizing agents. $NO_2^\bullet$ may induce peroxidation of lipids in the lung by abstracting hydrogen from polyunsaturated fatty acid side chains $$NO_2^\bullet + lipid-H \rightarrow HNO_2 + lipid^\bullet$$ O_3 can oxidize lipids and proteins directly. Sulphur dioxide (SO_2) toxicity has been *suggested* to involve free radicals.
Paraquat	Redox-cycling herbicide (Chapter 5, Table 3). Severe lung damage is probably due to excess $O_2^{\bullet-}$ formation *in vivo*.

Table 2. *Some disorders to which oxidative damage may contribute significantly*

Ionizing-radiation-induced tissue injury

Consequences of chronic selenium deprivation

Consequences of inborn defects in antioxidant defence enzymes[†]

Neurological disorders caused by chronic vitamin E deprivation (in diseases affecting intestinal fat absorption)

Retinopathy of prematurity (retrolental fibroplasia)

Tissue injury in copper-overload disease (Wilson's disease)

Tissue injury in iron-overload diseases (idiopathic haemochromatosis, thalassaemia)[*]

Cataract induced by ionizing or ultraviolet radiation

Consequences of exposure to elevated O_2 levels

Ultraviolet-light-induced skin injury

[†] Inborn deficiencies in glutathione reductase, glutathione peroxidase, glutathione-synthesizing enzymes, or the enzymes that make NADPH (needed for glutathione reductase) in cells can produce severe symptoms, depending on the degree of deficiency. So can deficiencies in superoxide dismutase; there is a partial defect in SOD in one form of the crippling neurodegenerative disease *amyotrophic lateral sclerosis* (Lou Gehrig's disease). By contrast, an *excess* of Cu,Zn-SOD has been suggested to contribute to the pathology of Down's syndrome (Chapter 3).

[*] Thalassaemias are inborn defects in haemoglobin synthesis. Patients are kept alive by regular blood transfusions, but each unit of blood contains about 0.2 g of iron. Iron overload eventually results; to prevent this, patients are given the iron chelator *desferrioxamine*.

radicals are also probably involved in the consequences of exposures to elevated O_2 concentrations, to certain toxins (Table 1), and in diseases that result in deficiencies of antioxidant defence enzymes or increases in the amounts of iron and copper in the body (Table 2).

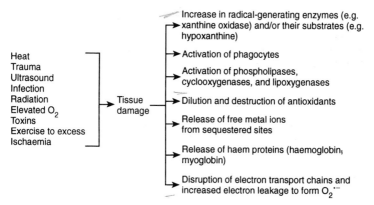

Figure 1. Some of the mechanisms by which tissue damage can cause oxidative stress.

Tissue injury as a cause of oxidative stress

Although oxidative stress may play an important role in some human disease, **it is probably safe to say that most (and perhaps all) forms of tissue injury themselves *lead* to oxidative stress.** Figure 1 summarizes some of the reasons for this.

When body tissues are damaged, neutrophils and other phagocytes enter the damaged area (see the Appendix to Chapter 5). These cells become activated, releasing $O_2^{\cdot-}$ and HOCl, that can help to remove foreign organisms at the injury site. However, this also imposes a stress on the surrounding tissues. For example, HOCl can oxidize thiol (−SH) groups on proteins. When tissues are crushed or torn, iron can be released from cells, both as free iron and in the form of haem-containing proteins (such as myoglobin), and as the storage protein ferritin. Free iron can convert $O_2^{\cdot-}$ and H_2O_2 into highly damaging $OH^{\cdot}$; $O_2^{\cdot-}$ can cause a limited release of iron from ferritin, and H_2O_2 can release iron from haem proteins. Haem proteins can accelerate lipid peroxidation. Since the iron content of human tissues increases with age, perhaps iron is mobilized in greater amounts as a result of injury and free radical damage becomes more severe in injured older tissues.

Table 3. *Some examples of ischaemic injury in human disease*

Stroke

Myocardial infarction, angina pectoris*

Reattachment of severed limbs

Frostbite

Shock due to excessive blood loss (O_2 supply to many body tissues is impaired)

Shock due to other causes (shock results in low blood pressure, which cuts O_2 supply to many tissues)

Severe crush injury

The tissues are deprived of O_2 (*ischaemia*). This will eventually kill them. If reoxygenation occurs before the tissue is dead, a reoxygenation injury may be important in worsening the effects of the ischaemia.

* Angina pectoris is chest pain arising when the amount of blood flowing through the coronary arteries is decreased because of atherosclerosis (Chapter 7). When the heart rate increases (e.g. by exercise or stress), the heart muscle outruns its limited blood supply and becomes transiently ischaemic. It has been suggested that some degree of reperfusion injury results when the heart rate returns to normal.

Ischaemia-reoxygenation

An apparently paradoxical cause of oxidative stress in human disease is *ischaemia-reoxygenation* (Table 3). If body tissues are deprived of O_2, they are injured and will eventually die. This is because O_2 is needed for mitochondria to make the cell's energy currency (ATP) and mitochondria are the major source of ATP in most human cells. One example of this *ischaemic* injury is the tissue damage resulting from application of a tourniquet to an injured limb for too long, cutting off the arterial blood supply. *Stroke* and *myocardial infarction* are other examples. In stroke, a blood clot forms in an artery in the brain, shutting off the blood supply to part of this organ. In a myocardial infarction (heart attack), a clot forms in one of the coronary arteries that supply the heart itself with blood.

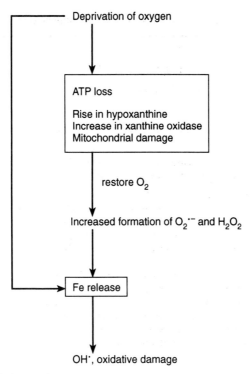

Figure 2. Ischaemia/reoxygenation.

Deprivation of blood flow, and the resulting deprivation of O_2 cause many metabolic changes. Ischaemic tissues become more acidic (their pH falls), ATP levels drop, and iron ions are released from storage sites. Mitochondria may be disrupted. One change that has attracted much attention in the medical literature is that ischaemic tissues increase their levels of *xanthine oxidase*, an enzyme that uses O_2 to oxidize xanthine and hypoxanthine into uric acid. The O_2 is converted into both $O_2^{\cdot-}$ and H_2O_2 during this process (see Chapter 2). Hypoxanthine also accumulates in ischaemic tissues (Figure 2) because it is made as ATP levels drop.

When a tissue is ischaemic, the O_2 supply must be restored as fast as possible before the cells die. For example, clot-dissolving agents such as *streptokinase* may be injected into patients who have suffered a heart attack. However, the restoration of O_2 (although essential) produces an oxidative stress upon the tissue: this is the paradox of *reoxygenation injury*. Damaged mitochondria may be 'leakier' than normal, so that, when they start working again, more electrons escape from the correct path in the electron transport chain and form $O_2^{\cdot-}$. Once O_2 is available, xanthine oxidase can oxidize the accumulated hypoxanthine, making $O_2^{\cdot-}$ and H_2O_2. If iron has been released within the tissue, $OH^{\cdot}$ may form and worsen the damage (Figure 2). The paradox is that, although restoring the O_2 is essential, the restoration adds an additional insult to the tissue in the form of oxidative stress. Reoxygenation injury makes little difference if the tissue has already been killed by the ischaemia but may be a significant additional source of injury after brief periods of O_2 deprivation. Another problem is that reoxygenation of badly damaged ischaemic tissue can wash metal ions, xanthine oxidase, and other potential toxins into the bloodstream and injure other body tissues (Table 3).

If tissue injury leads to oxidative stress (Figure 1), does this then make a major additional contribution to the injury, or is it insignificant? The answer probably differs in different diseases

Table 4. *Some disease in which oxidative stress is a consequence of the disease but might make some contribution to tissue injury*

Atherosclerosis (see Chapter 7)

Stroke (if reperfusion occurs before irreversible injury to tissue)

Myocardial infarction (if reperfusion occurs before irreversible tissue injury)

Rheumatoid arthritis

Inflammatory bowel diseases (Crohn's disease, ulcerative colitis)

Shock

Traumatic injury to brain and spinal cord

(Table 4). Several diseases in which an important role for oxidative stress is envisaged will now be briefly discussed.

Rheumatoid arthritis and other chronic inflammations

Rheumatoid arthritis is a chronic inflammation of the joints, producing painful swelling and loss of mobility. It may result in disabling destruction of the joint. The rheumatoid joint is a site of intense oxidative stress; for reasons unknown, large numbers of macrophages and neutrophils are present, releasing $O_2^{\cdot -}$, H_2O_2, HOCl, $NO^{\cdot}$, and other potentially damaging products. Bleeding within the joint can raise its iron content, allowing $OH^{\cdot}$ formation. A comparable situation exists in the inflammatory bowel diseases, in which there is excessive accumulation and activation of phagocytes in the gut, causing severe damage. *Ulcerative colitis* is one example.

Brain and spinal cord injury

In the test tube, damaged brain undergoes lipid peroxidation very fast, one reason being that iron is very easily released from disrupted brain cells. In addition, brain tissue is very rich in polyunsaturated fatty acid side chains, the levels of antioxidant defence enzymes in the brain are only moderate, and the fluid surrounding the brain (cerebrospinal fluid) has little transferrin and so cannot bind released iron. Free radical reactions probably occur in damaged brain tissue after injury (e.g. that caused by a blow to the head). It has been suggested that these reactions worsen the consequences of initial injury to the brain or spinal cord, by spreading the damage into surrounding areas. Similar free radical reactions may occur after stroke, spreading the damage beyond the ischaemic area.

Adult respiratory distress syndrome (ARDS)

ARDS, acute respiratory failure due to clogging of the lungs with fluid (*oedema*), can arise in patients with several different serious problems, including major infections, serious injury, severe shock, extensive burns, and inhaling the contents of the stomach into the lungs. The hallmark of ARDS is an accumulation of large numbers of neutrophils in the lung, where they are thought to become activated and produce $O_2{}^{\cdot -}$, H_2O_2 and $HOCl$. It is widely believed that this oxidative stress contributes to the lung injury and oedema in ARDS. Although the occurrence of oxidative stress has been demonstrated in ARDS patients, there is as yet no direct *evidence* that it is a major contributor to the lung injury.

Eye injury
The development of opacity of the lens of the eye (*cataract*) in old people involves oxidation and cross-linking of lens proteins. Its development can be accelerated, presumably by free radical mechanisms, as a result of exposure to ionizing radiation or to ultraviolet light. Bleeding within the eye or the accidental introduction of an iron object into the eye can cause severe damage, in which iron-dependent free radical damage has been implicated. For example, the retina is rich in polyunsaturated fatty acid side chains and appears prone to undergo lipid peroxidation.

How can oxidative stress be treated?
Oxidative stress is probably an inevitable accompaniment of disease (Figure 1). In some diseases, it may make a significant contribution to the disease pathology (Table 4), whereas in others (perhaps most) it does not. How can disease-related oxidative stress be dealt with?

Eating a diet rich in fruits and vegetables will ensure adequate levels of antioxidant nutrients in the tissues and help the body to resist disease-related oxidative stress. If, for example, levels of vitamin E in the brain are subnormal, the consequences of a stroke, or of traumatic injury, may well be more severe because uncontrolled lipid peroxidation could spread the injury to surrounding areas.

In general, trials of antioxidants in the treatment of human disease have given unimpressive results to date, except in diseases in which oxidative stress may be causative (e.g. retinopathy of prematurity, haemolytic syndrome in premature babies, neurological degeneration caused by tocopherol deficiency in patients unable to transport fats through the gut, and Keshan disease). Several reasons can account for this:

(1) Oxidative stress may occur, but be unimportant in the pathology of that particular disease.
(2) Insufficient antioxidant has been used to reach the site at which it is needed or to remove enough radicals.
(3) The wrong antioxidant has been used. Oxidative stress affects many different process (see Chapter 4). If, in a particular situation, direct free radical damage to proteins or to DNA was the important injury mechanism, then an inhibitor of lipid peroxidation is unlikely to be protective.
(4) Antioxidant has not been administered for long enough.

Fortunately, several well designed clinical trials are now in progress to test the effects of antioxidants against atherosclerosis (see Chapter 7) and degenerative brain diseases, in ameliorating the consequences of stroke and traumatic brain injury, in ARDS, and in HIV infection (does correcting the apparent deficiency in GSH give any clinical benefit?). The results are awaited with interest.

The ageing process: are free radicals involved?

All would live long, but none would be old.
                                  Poor Richard Benjamin Franklin (1749)

We are all ageing and will eventually die. The maximum human life span is probably around 110–120 years, but few people in the third world achieve it because of early death due to malnutrition or infection. In the USA and Europe, such deaths are rare. The few people who die under age 35 usually do so as a result of accidents, although deaths from AIDS are becoming significant. Older people in Europe and the USA more often die of myocardial infarction or cancer. The incidence of cancer varies strikingly with age. A few cancers occur in young people, such as cancer of the testis and some types of leukaemia. For most cancers, however, the rate of incidence rises very sharply with age (Figure 3), that is cancer is an age-related disease. It takes many years to accumulate all the different changes in DNA that are needed for most cancers to develop.

The reasons for ageing have been the subject of considerable

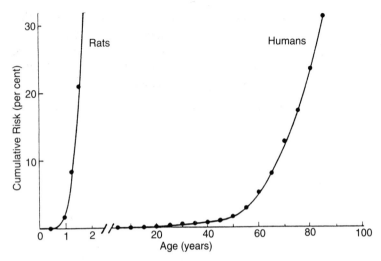

Figure 3. The rise in cancer incidence with age, showing the cumulative net risk of death from cancer in humans. Note the very sharp rise in cancer incidence with age. Apart from a few rare cancers, cancer is generally a disease of older people. The same phenomenon occurs, on a much shorter time-scale, in other mammals such as rats. (From *Free Radical Research Communications*, **7**, 122, by courtesy of Professor Bruce Ames and Harwood Academic Press.)

speculation, but there is little or no hard information. Many years ago, it was realized that the basal rate of metabolism of animals (i.e. how fast the cells are working when the body is at rest) is approximately inversely related to their life span; in general, larger animals consume less O_2 per unit of body mass than do smaller ones, and they live longer. These data suggest that ageing is somehow related to O_2 metabolism, and there have been many speculations that free radicals are involved. Thus, the faster O_2 is consumed by an organism, the more free radicals it might make.

However, there is no convincing *evidence* that free radicals *cause* the ageing process. Antioxidant defences do not appear to fail with age, nor is there any great accumulation of oxidatively damaged molecules. Feeding antioxidants to mammals in the laboratory has never been convincingly shown to make them live longer. As one ages, tissues deteriorate, and this could lead to secondary free radical damage, as might all forms of tissue injury (Figure 1). Thus, claims that antioxidants, royal jelly, or any other potion will make you live longer are without scientific foundation.

7 Free radicals and cardiovascular disease

What you eat, you are.

The Beatles, George Harrison (1968)

In our preoccupation with the complexities of the scientific evidence, we tend to forget that the best dietary advice can be simple: eat a varied diet, not too much of anything, and enjoy it.

Michael Gurr (1992)

Coronary heart disease is a major cause of death in the USA and Europe. A heart attack (*myocardial infarction*) is usually the consequence of two events—the narrowing of coronary arteries by *atherosclerosis* (Figure 1) and the formation of a blood clot (*thrombus*) in a narrowed artery, which blocks it completely and renders a portion of the heart muscle ischaemic (lacking in O_2). Atherosclerosis of arteries in the brain also predisposes to stroke. There have been many speculations that dietary intakes of polyunsaturated fatty acids, cholesterol, antioxidant nutrients, iron, copper, and selenium affect the development of atherosclerosis. The purpose of this chapter is to review what is, and is not, actually known about this complex area.

The nature of atherosclerosis

Atherosclerosis first appears in arteries as *fatty streaks*. These are slightly raised, yellow, narrow areas of abnormality on the arterial

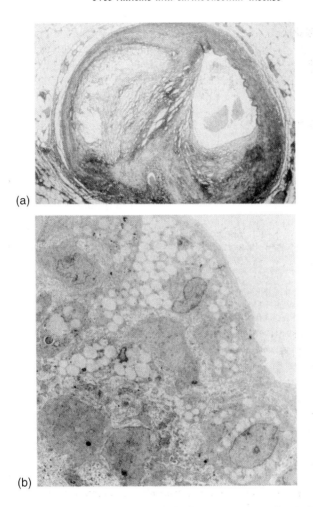

(a)

(b)

Figure 1. Microscopic appearance of sections through a human coronary artery with severe atherosclerosis. (a) Heavily atherosclerosed artery. (b) Section through a lesion showing the cells types present. Note numerous lipid-laden foam cells, many of which are produced when macrophages ingest peroxidized lipids. (Courtesy of Upjohn Co., 'Current Concepts' series.)

wall. Under the microscope, they are seen to contain *foam cells*, cells that have accumulated large amounts of lipid (Figure 1). Many foam cells arise from *macrophages* (see the Appendix to Chapter 4). Fatty streaks sometimes develop into *fibrous plaques*, which are rounded raised lesions, usually off-white in colour and perhaps a centimetre in diameter. A typical fibrous plaque has a fibrous cap covering an area rich in macrophages, lymphocytes, and muscle cells, perhaps with a core containing dead cells and cholesterol crystals. *Complicated plaques* may arise from fibrous plaques by deposition of calcium, thrombosis, and other events. It is often the rupture of the cap of a plaque and release of its noxious contents that triggers the thrombus formation that finally blocks an artery and produces ischaemia.

Fatty streaks develop very early in life. Most children in the USA and in Europe have them. In the disease *familial hypercholesterolaemia* the accumulation of plaques is much increased and children as young as two can suffer heart attacks. Familial hypercholesterolaemia is a disease in which the body cannot handle low-density lipoproteins (LDLs) properly. As explained in Chapter 3 (Table 2), LDLs are lipid–protein complexes in the blood plasma that are rich in cholesterol and cholesterol esters, and serve to supply cholesterol to cells that need it. The LDLs bind to *receptor proteins* on the surface of cholesterol-requiring cells and enter the cells. In familial hypercholesterolaemia, the receptors are defective, so that blood LDL (and cholesterol) levels become abnormally high.

The consequences of familial hypercholesterolaemia have focused attention on high blood cholesterol levels as a risk factor for atherosclerosis. Remember, however, that cholesterol (in normal amounts) is very useful in the body (see the Appendix to Chapter 1), being essential for membrane structure and synthesis of several compounds. If not enough cholesterol is eaten, the liver will make it from other dietary fats or from sugars.

Thrombosis

When body tissues are damaged, various clotting factors are released into the blood, which stimulate a series of chemical changes in blood proteins. At the end of this sequence of events, a soluble blood protein (*fibrinogen*) is converted into an insoluble protein that forms strands (*fibrin*). The small blood cells known as *platelets* stick to damaged blood vessel walls and aggregate together. Fibrin, platelets, and trapped red blood cells form the thrombus. The tendency of blood to clot depends on the activity of the various clotting proteins, the blood fibrinogen concentration, and upon how 'sticky' the platelets are.

Risk factors for myocardial infarction

The risk of myocardial infarction can be raised by facilitating atherosclerosis, thrombosis, or both. No one is really sure what is the precise sequence of events. However, certain factors increase the risk that an individual will suffer a heart attack. The best-established risk factors are

- increasing age;
- family history of heart disease;
- high blood pressure;
- male gender (up to age 45, men are much more prone to heart attacks than women; after the menopause, the difference gradually disappears);
- cigarette smoking;
- high plasma cholesterol, particularly cholesterol in LDLs (levels of total plasma cholesterol above 240 mg/decilitre (6.2 mM) are considered significantly elevated);
- low blood levels of vitamins C and E.

Risk factors are cumulative—the more you have, the greater your risk. Remember, however, that these are *correlations*; they do not mean that the factors are direct causes of heart attacks. Thus, several studies have shown that low blood levels of antioxidant vitamins (C and E) are associated with increased risk of heart disease (e.g. see the Appendix to Chapter 4). Low blood levels of these vitamins signify a diet poor in fruits and vegetables. Is it the lack of E and C that is important, or is it some other constituents of these foods, or is it both factors? Is the 'advantage' of women over men in terms of lower risk of heart disease due to different sex hormones, lower body iron stores (see Chapter 4, Figure 2), differences in blood lipids, or other factors? Does cigarette smoke predispose to athero-sclerosis directly, or does it act by raising plasma cholesterol, encouraging blood clotting, depleting antioxidant vitamins or is it a 'marker' of the fact that heavy cigarette smokers often eat a poor diet? Other suggested risk factors include being overweight, lack of exercise, stress, too much alcohol, high levels of fibrinogen (a blood clotting protein) in the blood and diabetes.

Dietary fats and heart disease

It has been widely believed by nutritionists that a high intake of fat in the diet raises plasma cholesterol levels and increases the risk of heart disease. This is sometimes call the 'lipid hypothesis' of coronary heart disease. In adults with abnormally high plasma cholesterol (≥ 6 mM), cutting the amount of total fat and/or the amount of cholesterol-rich foods in the diet does decrease plasma cholesterol somewhat (Table 1 lists some foods that might be avoided). However, it has proved difficult to produce much of a decrease in humans who have cholesterol levels in the normal US or UK range, by dietary changes. In the UK, average fat intake is about 100 g per day, providing about 42 per cent of energy requirements. Cholesterol in the diet falls in the range of 270–750 mg per day and comes only from foods of animal origin. Not all of it is necessarily absorbed. Because the body can make cholesterol, for most people with normal blood cholesterol levels, restriction of dietary choles-terol has only small effects on the blood level. It can be beneficial in subjects with elevated (>6 mM) blood cholesterol. In patients with elevated plasma cholesterol, drugs (such as *cholestyramine*, *simvastatin*, and *lovastatin*) are available that decrease its level. Clinical trials have shown some effects of such drugs in decreasing mortality from heart disease, but the results are not very striking. Hence, many experts are now sceptical about whether changes in diet or the taking of drugs to decrease cholesterol levels are really justified, except in patients with plasma cholesterol levels greater than about 6 mM.

Table 1. *Dietary sources rich in cholesterol. In the United Kingdom, average fat intake is about 100 g/day, providing about 42% of energy requirements. Cholesterol in the diet falls in the range of 270–750 mg/day and comes only from foods of animal origin. Not all of it is necessarily absorbed. Because the body can make cholesterol, for most people with normal blood cholesterol levels, restriction of dietary cholesterol has only small effects on the blood level. It can be beneficial in subjects with elevated (>6 mM) blood cholesterol*

Food	Cholesterol (mg/100 g)
Eggs (all cholesterol is in the yolk)	385
Milk	14
Double cream	140
Single cream	60
Cheese	100
Cooked meat (beef, pork, lamb)	80–110
Prawns	200
Cooked liver	~400
Cooked kidney	~600
Fish	60–100
Fruit, grains, vegetables	0

Polyunsaturated fats

Another frequent proposal has been that we should switch from saturated fats in the diet to polyunsaturated fats. Such dietary changes have been shown experimentally to lead to decreases in plasma levels of LDL cholesterol, although there is considerable variation in the size of the decrease in different people. However, not all saturated fatty acids in lipids raise plasma cholesterol: only $C_{12:0}$ (lauric), $C_{14:0}$ (myristic), and $C_{16:0}$ (palmitic) acids seem to be important. Monounsaturates such as oleic acid ($C_{18:1}$) seem to decrease plasma cholesterol as effectively as do polyunsaturates. The saturated fatty acid stearic acid ($C_{18:0}$) is easily converted into oleic acid in humans, which perhaps explains reports that stearic acid might actually lower plasma cholesterol. Thus butter fat, which contains about 30 per of its fatty acids as monounsaturates, may not be as bad for us as some nutritionists have suggested, despite its 30 per cent content of palmitic acid side chains.

One particular worry that has been raised about polyunsaturates is their tendency to undergo lipid peroxidation after free radical attack, a tendency not shown by saturated fatty acids and less shown by monounsaturated fatty acids (monounsaturated fats such as oleic acid *can* be peroxidized, however: for example, olive oil can go rancid if stored too long or mishandled). Another factor to be considered is the effect of lipids on how easily the blood clots. For example, diets rich in fish oils (as eaten by Eskimos) seem to deter blood clotting somewhat, and so might diminish thrombus formation and heart attacks irrespective of their effect (if any) on atherosclerosis.

Free radicals and atherosclerosis

The origin of atherosclerosis is uncertain, but it is widely believed that it begins as a result of damage to the *vascular endothelium*, the continuous layer of cells that lines the inner wall of blood vessels. These cells form a barrier between the blood and the deeper layers of the blood vessels, but they have a multiplicity of other important functions. They synthesize many chemicals that regulate the activity of platelets and other aspects of thrombosis. By forming certain proteins on their surface, endothelial cells can cause phagocytes in the blood to adhere and begin their journey through the vessel wall into the tissues at sites of injury (see the Appendix to Chapter 4). Endothelial cells convert the amino acid L-arginine into the free radical nitric oxide, $NO^{\bullet}$, which acts on the muscle in the vessel walls to relax it and lower blood pressure. It has even been suggested that endothelium may sometimes make superoxide radicals ($O_2^{\bullet-}$), perhaps using the enzyme xanthine oxidase, and that this $O_2^{\bullet-}$ is involved in adherence of phagocytes.

Damage to the vascular endothelium, and initiation of atherosclerosis, might happen in several ways. High blood pressure could cause turbulent blood flow in some vessels, causing mechanical damage. Some chemicals in the blood (e.g. toxins absorbed from inhaled cigarette smoke or the high blood glucose concentrations in patients with diabetes) might injure the endothelium. In patients with the inborn disease *homocystinuria*, high blood levels of the compound *homocysteine* may damage the endothelium.

Damage to the endothelium appears to cause attachment of the phagocytes known as *monocytes* from the blood (see the Appendix to Chapter 4). The monocytes migrate into the vessel wall and develop into macrophages. Both cell types secrete $O_2^{\cdot-}$ and H_2O_2 when activated. Somehow, these various events in the vessel wall create a localized oxidative stress. When LDLs from the plasma enter such regions in the vessel wall, they can be subjected to oxidative stress and undergo lipid peroxidation (precise details of how this happens are unclear). Peroxidation will be facilitated if the LDLs are rich in polyunsaturated fatty acid side chains and low in antioxidants. LDLs rich in saturated and monounsaturated fatty acid side chains and in antioxidants (such as vitamin E) would be expected to be more resistant to peroxidation (see Chapter 4). Another important factor influencing these events might be the levels of vitamin C in the vessel wall, since Vitamin C can regenerate vitamin E from its radical form within LDLs (see Chapter 3, Figure 5).

Peroxidized LDLs may be toxic to endothelium, worsening the damage and attracting more monocytes. Peroxidation generates phospholipid hydroperoxides and also cholesterol oxidation products, both of which are noxious. Peroxidized LDLs might also irritate macrophages, causing them to secrete growth factors that encourage proliferation of muscle cells, and to produce chemical 'signal molecules' that attract lymphocytes into the lesion. Peroxides in LDLs can decompose (in reactions catalysed by free iron or copper ions) into a wide range of noxious products, including lipid free radicals and aldehydes. When formed in LDLs, these radicals and aldehydes can attack the protein in LDLs. LDLs with protein modified in this way are recognized by special proteins (*scavenger receptors*) on the surface of macrophages. Scavenger receptors cause the uptake of oxidized LDLs into macrophages at a very fast rate. The oxidized LDL is removed from the environment by the macrophage, and lipid-laden foam cells (Figure 1) are generated.

Atherosclerosis and antioxidants: a key relationship?

It is interesting to note how many of the risk factors for atherosclerosis could be explained by a key role for lipid peroxidation in
the development of atherosclerosis:

(1) High plasma cholesterol usually means that a lot of LDL is
    available for peroxidation: cholesterol oxidation products in the
    vessel wall might also be toxic.

(2) Smoking can raise LDL levels and deplete vitamins C and E in
    the body. Perhaps components of inhaled smoke can enter the
    blood and help LDLs to oxidize.

(3) Low plasma levels of antioxidant vitamins mean that LDLs
    would be less resistant to peroxidation.

(4) A high intake of monounsaturated fats might decrease per-
    oxidation of LDLs, since such fats are more resistant to
    peroxidation.

(5) The complex effects of intake of dietary polyunsaturated fats
    upon the incidence of heart disease may be a balance between
    their tendency to lower plasma cholesterol and to make LDLs
    more prone to peroxidation. However, remember that many
    foods rich in polyunsaturates are also rich in vitamin E, provided
    that they have not been stored too long or mishandled.

The uptake of oxidized LDL by macrophages to produce foam cells
might have evolved as a protective mechanism for removing the
cytotoxic LDL. It could also, however, cause the macrophages to
secrete chemical factors that stimulate development of the lesion.
Paradoxically, therefore, the LDL that has begun peroxidation in the
vessel wall but has not yet been changed enough to be recognized by
the scavenger receptors of macrophages might be ultimately more
dangerous.

What evidence is currently available concerning the proposal that
lipid peroxidation is important in the development of atherosclerosis
and what are the weak links in it? First, it has been possible to
demonstrate oxidized LDLs in human atherosclerotic lesions (and in
similar lesions from animals). However, remember that the lesion is
a site of tissue injury, and injury tends to produce additional

damage due to free radicals (see Chapter 6, Figure 1). Demonstration of lipid peroxidation is not evidence that lipid peroxidation plays an important role in atherosclerosis.

Second, antioxidants have been shown to diminish the incidence of atherosclerosis in rabbits fed diets very rich in cholesterol. However, the very high plasma cholesterol levels achieved in these experiments may be irrelevant to the development of atherosclerosis in most humans (with the exception of familial hypercholesterol-aemia). The antioxidants shown to be effective, probucol, and the food antioxidant BHT (butylated hydroxytoluene) might have other effects on the body. Thus probucol lowers HDL and LDL levels as well as acting as a chain-breaking antioxidant. (Probucol acts like vitamin E and BHT in scavenging peroxyl radicals, so halting the chain reaction of lipid peroxidation.)

Third, the proposal could explain why plasma levels of certain antioxidant vitamins are inversely related to the incidence of heart disease. Caution is required however, since plasma levels of these vitamins are a 'marker' of a diet rich in fruit and vegetables, and it could be many other factor or combination of factors in that diet that are protective. Indeed, we have already seen (Chapter 4, Figure 5) that the sensitivity of LDLs isolated from different human subjects to free radical damage in the test tube does not correlate with the LDL vitamin E content. Epidemiological studies show an inverse relation between heart disease and β-carotene levels, yet loading LDLs with β-carotene does not make them more resistant to peroxidation. The real importance of antioxidants in preventing atherosclerosis can only be established by carefully controlled clinical trials in which volunteers are supplemented with individual antioxidants.

Another suggestion is that antioxidants might affect thrombosis. Thus it has been hypothesized that deficiencies in vitamin E could affect endothelial and platelet function and increase the tendency to thrombosis.

Exercise

It is widely accepted that regular exercise is 'good for you' and helps to prevent heart disease. In addition to 'toning up' the heart and other muscles, several studies suggest that exercise can raise the proportion of plasma cholesterol in HDL, reportedly a protective agent against atherosclerosis. However, severe exercise in untrained individuals produces tissue injury, which would be expected to lead to oxidative stress (see Chapter 6, Figure 1). Indeed, oxidative damage has been reported to result from severe forced physical exercise in rats. Experiments on animals, however, suggest that by carefully training, antioxidant defences increase and decreased tissue injury and oxidative damage is observed resulting in an overall beneficial effect.

When athletes train, several physiological responses occur. These include falls in plasma iron and zinc, and a rise in plasma copper. Because of these changes it is often said that athletes are 'deficient' in iron ('sports anaemia') and zinc but this is rarely true (except

Table 2. *The authors' dietary recommendations*

Eat plenty of fresh fruit and vegetables, at least five different portions per day.

Minimize intake of fats and red meats, but do not become paranoid about it. Don't worry about polyunsaturates versus saturates.

Check your cholesterol level. If 200 mg/100 ml or below don't worry. If at or above 250 mg/100 ml seek medical advice.

Consume no more than 300 units (200 mg) of vitamin E (d-α-tocopherol, *not* dl-α-tocopherol) per day from a reliable source such as the 'own brand' of a reputable chain drug store. Take with food as you need fat to absorb it.

If you wish, consume up to 250 mg of vitamin C per day. Again, select a reputable supplier (e.g. the 'own brand' of a reputable chain drug-store). If you smoke, stop. If you can't, eat plenty of fruits and vegetables and consider supplementing with more vitamin C (Table 2, p. 67).

Do not take any form of iron supplements unless there is a clearly identified medical need monitored by laboratory tests.

We see no case at present for consuming β-carotene supplements.

---

possibly for iron in female athletes). Indeed, an additional reason suggested for the apparent protection achieved by regular exercise against cardiovascular disease is that it tends to lower body iron stores.

Authors' conclusion

Antioxidants are clearly important to human life, but they are not elixirs of life. What sense can be made of all the various dietary recommendations? In strict scientific terms, little or nothing is proved. However, the authors have, from many years of research and listening to other experts, formed their own opinions and applied them to their own lifestyles (Table 2). We do not recommend what we do to others, who must make up their own minds from the data presented! We hope that this little book will assist readers in the decision process.

Further reading

Ames, B.N. (1989). Endogenous oxidative DNA damage, ageing, and cancer. *Free Radical Research Communications*, **7**, 121–8.

Ashwell, M. (ed) Diet and Heart Disease. A Round Table of factors. British Nutrition Foundation, London, 1993.

Balentine, J.D. (1982). *Pathology of oxygen toxicity*. Academic Press, New York and London.

Bien, J.G. (1989). Are the recommended allowances for dietary antioxidants adequate? *Free Radical Biology and Medicine*, **3**, 193–7.

Block, G. (1991). Epidemiological evidence regarding vitamin C and cancer. *American Journal of Clinical Nutrition*, **54**, 1310S–14S.

Canfield, L.M. *et al.* (1992). Carotenoids as cellular antioxidants. *Proceedings of the Society of Experimental Biology and Medicine*, **200**, 260–5.

Crawford, M.A. (1992). The role of dietary fatty acids in biology: their place in the evolution of the human brain. *Nutrition Reviews* **50**, 3–11.

Dietary reference values for food energy and nutrients for the United Kingdom (1991). Report on health and social subjects, 41. Her Majesty's Stationery Office, London.

Esterbauer, H. *et al.* (1991). Effect of antioxidants on oxidative modification of LDL. *Annals of Medicine*, **23**, 573–81.

Gray, J. and Buttriss, J.L. (1992). Coronary heart disease II, Fact File 8, National Dairy Council, London.

Gutteridge, J.M.C. and Halliwell, B. (1989). Iron toxicity and oxygen radicals. In *Iron chelating therapy*, Vol. 2 (ed. C. Hershko), pp. 195–256. Bailliere Tindall, London.

Hall, E.D. and Braughler, J.M. (1989). Central nervous system trauma and stroke II. Physiological and pharmacological evidence for involvement of oxygen radicals and lipid peroxidation. *Free Radical Biology and Medicine*, **6**, 303–13.

Halliwell, B. and Gutteridge, J.M.C. (1989). *Free radicals in biology and medicine*, 2nd edn. Clarendon Press, Oxford.

Halliwell, B. and Gutteridge, J.M.C. (1990). The antioxidants of human extracellular fluids. *Archives of Biochemistry and Biophysics*, **280**, 1–8.

Halliwell, B., Gutteridge, J.M.C. and Cross, C.E. (1992). Free radicals and human disease—where are we now? *Journal of Laboratory and Clinical Medicine*, **119**, 598–620.

Harman, D. (1993). Free radical involvement in ageing. Pathophysiology and therapeutic implications. *Drugs and Ageing*, **3**, 60–80.

Janero, D.R. (1991). Therapeutic potential of vitamin E in the pathogenesis of spontaneous atherosclerosis. *Free Radical Biology and Medicine*, **11**, 129–44.

Kappus, H. and Diplock, A.T. (1991). Tolerance and safety of vitamin E: a toxicological position report. VERIS, LaGrange, Illinois.

Lauffer, R.B. (ed.) (1992). *Iron and human disease*. CRC Press, Boca Raton, Florida.

Mino, M. (1992). Clinical uses and abuses of vitamin E in children. *Proceedings of the Society of Experimental Biology and Medicine*, **200**, 266–70.

Norday, A. (1991). Is there a rational use for n–3 fatty acids (fish oils) in clinical medicine? *Drugs*, **42**, 331–42.

Orrenius, S. *et al.* (1988). Ca^{2+}-activated mechanisms in toxicity and programmed cell death. *ISI Atlas of Science, Pharmacology*, pp. 319–24.

Reed, D.J. (1990). Review of the current status of calcium and thiols in cellular injury. *Chemical Research in Toxicology*, **3**, 495–502.

Saltman, P., Gurin, J. and Mothner, I. (1987). *The California nutrition book*. Little, Brown and Company, Toronto.

Scandalios, J.G. (ed.) (1992). *The molecular biology of free radical scavenging systems*. Cold Spring Harbor Press, New York.

Sherman, A.R. (1992). Zinc, copper, and iron: nutriture and immunity. *Journal of Nutrition*, **122**, 604–9.

Sies, H. (ed). *Oxidative Stress*, 1st edn (1985) and 2nd edn (1991). Academic Press, New York and London.

Steinberg, D. (1993). Antioxidant vitamins and coronary heart disease. *New England Journal of Medicine*, **328**, 1487–9.

Steinberg, D. *et al.* (1991). Beyond cholesterol: modifications of low density lipoprotein that increase its atherogenicity. *New England Journal of Medicine*, **320**, 915–24.

Sullivan, J.L. (1988). The iron paradigm of ischaemic heart disease. *American Heart Journal*, **117**, 1177–88.

von Sonntag, C. (1987). *The chemical basis of radiation biology*. Taylor and Francis, London.

Taylor, A. (1993). Cataract: relationships between nutrition and oxidation. *Journal of the American College of Nutrition*, **12**, 138–46.

Ziegler, R.G. *et al.* (1992). Does ß-carotene explain why reduced cancer risk is associated with vegetable and fruit intake? *Cancer Research*, **52**, 2060S–6S.

Index